A GUIDE TO THE STUDY OF freshwater ecology

contours: studies of the environment

Series Editor
William A. Andrews
Associate Professor of Science Education
The Faculty of Education
University of Toronto

A Guide to the Study of ENVIRONMENTAL POLLUTION
A Guide to the Study of FRESHWATER ECOLOGY
A Guide to the Study of SOIL ECOLOGY
A Guide to the Study of TERRESTRIAL ECOLOGY

A GUIDE TO THE STUDY OF
freshwater ecology

Contributing Authors:
Daniel G. Stoker
Marcel Agsteribbe
Nancy R. Windsor
William A. Andrews

Editor:
William A. Andrews

Prentice-Hall of Canada, Ltd., Scarborough, Ontario

A Guide to the Study of
FRESHWATER ECOLOGY
© 1972 by W. A. Andrews
Published by Prentice-Hall of Canada, Ltd.,
Scarborough, Ontario.

Printed in Canada.
ISBN 0-13-370759-8   2 3 4 5 76 75 74 73 72

Prentice-Hall, Inc., *Englewood Cliffs, New Jersey*
Prentice-Hall International, Inc., *London*
Prentice-Hall of Australia, Pty., Ltd., *Sydney*
Prentice-Hall of India, Pvt., Ltd., *New Delhi*
Prentice-Hall of Japan, Inc., *Tokyo*

Design by Jerrold J. Stefl, cover illustration by Tom Daly,
text illustrations by James Loates.

PREFACE

Ecology is the study of the relationships between living things and their environments. This book is about freshwater ecology. It deals with the living things in and around ponds, lakes, rivers, and streams. It also explores the relationships between these living things and the environments in which they live.

Why do trout inhabit cool, rapidly flowing streams? Why do mosquitoes breed in stagnant waters? Why are leeches commonly found in sluggish streams? Why do crayfish prefer streams with rocky bottoms? In what ways do the chemical characteristics of the water determine which organisms can live in it? These are typical of the questions that you should be able to answer after you have studied the material in this book. The pursuit of answers to such questions should provide you with an exciting few weeks of laboratory and field work.

But this book will not have attained its main objective if you only learn how to *answer* such questions. Ultimately, you should understand the basic principles of ecology well enough that you can and will start *asking* questions: Why are trout no longer found in a once famous trout stream? What can be done to restore conditions favorable to trout? What will be the effects of damming up a river to provide an artificial lake for recreational purposes? What are the possible long-term effects of draining a marsh in a parkland area so picnickers will no longer be bothered by mosquitoes? Should insecticides be used in vacation areas to control insect pests? Do motorboats and their exhaust emissions affect aquatic life? Does it matter if they do? Is stocking a lake with young fish best answer to fishermen's complaints of poor catches? When you start asking questions like these, you will have become the type of citizen that is desperately needed to help restore a cleaner, more balanced environment for man as well as for all living things.

ACKNOWLEDGMENTS

This program was developed at the Faculty of Education, University of Toronto. The resources of the Faculty and the knowledge and skills of many student-teachers in the Environmental Studies option contributed greatly to the quality of the materials contained in this book and its companion volumes.

The authors are particularly appreciative of the competent professional help received from the Publisher. In particular, we wish to acknowledge the editorial assistance of Sue Barnes and Peter Anson. Their skill, knowledge, and patience are greatly appreciated. To Paul Hunt and Kelvin Kean we extend our sincere thanks for their help in planning this program in Environmental Studies. We are also grateful to Ron Decent and the many other members of the Prentice-Hall production staff for their effective work in the production of this book.

We wish also to thank Jim Loates for his excellent art work and Lois Andrews for her careful preparation of the manuscript.

D.G.S.
M.A.
N.R.W.
W.A.A.

CONTENTS

1

2

3

MAJOR FIELD STUDIES   134

6

CASE STUDIES   154

7

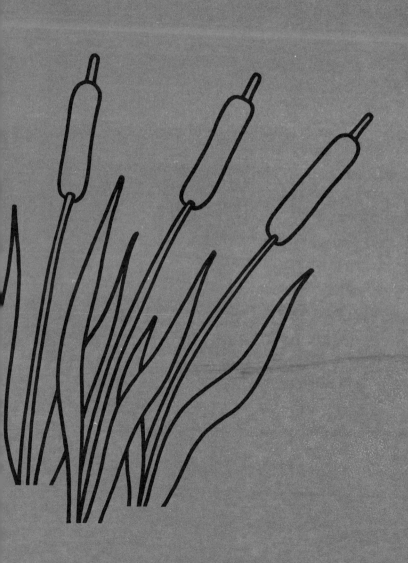

Introduction to Basic Principles

1

1.1   WHY STUDY FRESHWATER ECOLOGY?

Water, water, everywhere
Nor any drop to drink.

S. T. Coleridge, *The Rime of the Ancient Mariner*

Although this guide deals mainly with fresh waters, the above quotation is appropriate. Not many fresh waters are left today that are fit to be called fresh. The freshest water that you find may be as deadly a poison to wildlife as could ever be devised in a laboratory (Fig. 1.1). Therefore, even though pollution is not our main topic, some reference should be made to this most perplexing and dangerous of man's problems.

Too often, people become hysterical at the mention of words like pollution and eutrophication. They confuse these and other terms. In their concern with the moral, social, and political aspects of the problem, they may ignore its biological nature. This is not to underestimate the moral and social difficulties created by pollution, but the solution to our pollution problems requires more than condemnation of industries and governments. All of us are, in one way or another, responsible. The issue of pollution is much too important to be used as a political football.

What is needed then? Greater knowledge and understanding of ecology and environmental science is required of us all.

Fig. 1.1
Does this look like fresh water to you?

Only when we have this knowledge can the sources and effects of pollution be found, corrected, and avoided, with maximum efficiency and without needless emotional outbursts. In short, education is our most potent tool for solving our problems.

A few examples will illustrate this. Suppose that the residents of a town found that their once clean, clear stream became murky shortly after a factory had been built upstream. What would be the reaction? Probably a confrontation between the town and the factory. Yet the murky appearance might have been caused by a farmer who burned a few acres of shrub next to the river upstream. (Why would that make the water murky?) Or, imagine that you have a cottage on a lake and that every year one or two huge algal blooms make the lake unsightly and unsuitable for swimming. The lake could be very fertile you say. But why? Perhaps there are too many cottages and the lake simply cannot handle the nutrients that are dumped into it from septic systems. Perhaps the cause is a closed-down factory which for years poured its wastes into the lake. The excess nutrients in the lake could affect it for many more years to come.

Without suitable knowledge, we cannot even begin to find the causes of pollution, let alone correct them. Further, the problems are compounded by the fact that the natural increase of nutrients in waters (eutrophication) is nature's own way of polluting her waters. What then, is man-made pollution? According to the Environmental Pollution Panel of the President's Science Advisory Committee in its November, 1965 report, "Restoring the Quality of Our Environment":

> Environmental pollution is the unfavorable alteration of our surroundings, wholly or largely as a by-product of man's actions, through direct or indirect effects of changes in energy patterns, radiation levels, chemical and physical constitution and abundances of organisms. These changes may affect man directly, or through his supplies of water and of agricultural and other biological products, his physical objects or possessions, or his opportunities for recreation and appreciation of nature.

Study this definition carefully. What kind of knowledge do you require before you will be able to point your finger and say "That is pollution"?

Since all of man's pollution problems originate "wholly or largely as a by-product of man's actions," we need to study man *in relation to his environment*. It is not enough to study man alone, water alone, or plants alone. The study must relate to all living things and to their environments.

With this in mind, we will examine freshwater life and its habitat, to develop insight into the functional aspects of such smaller ecosystems (ecological systems). Many of the principles discovered by studying these ecosystems can be applied to the larger system of man and his environment.

You will need to learn specific methods and to use specific equipment in order to study particular aspects of aquatic ecosystems. Do not, however, lose sight of the whole, the complete system. As the saying goes, people sometimes fail to see the forest for the trees. The understandings that you gain from a specific field trip should be given wider applications. Man has polluted nearly every aspect of his environment—his water, his air, his land. He has filled his eyes, nose, lungs, and ears with undesirables. His actions show an indifference to the future. People like yourself, once aware of the basic principles of ecology, can give the future a chance. It is surprising that our environment has not already heaved its last polluted gasp, leaving us, the holy dumpers, holding the wrong end of the garbage bag! It is therefore essential that you, as students of ecosystems, bear in mind that what you do or do not do in the field may have a great bearing on not only the lesser system under study, but also the larger world-wide one, the biosphere. *Tamper with your study area as little as possible.* You can enjoy and learn on an outing without upsetting animal habitats, without damaging plant communities, and without collecting specimens that are not readily replaced.

This book has two major purposes. First we hope that you will develop sufficient interest and gain sufficient knowledge to be an informed as well as a concerned citizen when you are confronted with environmental problems that require attention. Second, we hope that your experiences here will open up a whole new world to you, the world of freshwater organisms. If the same thing happens to you as happened to the authors, your enjoyment of camping trips, hikes, and other outings will be greatly increased after you have learned some freshwater ecology. Aquatic plants and animals abound in shallow freshwater habitats, and all that is needed to discover this new world is a keen pair of eyes and a spirit of adventure. Have you ever seen the predaceous diving beetle larva lock its jaw around a graceful fairy shrimp, piercing its outer skeleton to suck out the body fluids? Have you observed the dragonfly nymph, one of the fiercest predators of the pond, fall victim to the beak of a giant water bug? Countless encounters and battles occur every minute in the underwater world. Even the antics of a slow-moving vegetarian like the caddisfly larva, camouflaged in its portable home of twigs, can captivate your attention for hours.

At first, the strangeness of aquatic life may be overwhelming. After all, you may be seeing everything for the first time. How can you expect to know the names, let alone the life cycles and habits of such bizarre creatures? A scientist who looks into a new world or field is faced with the same problem. Do laws of nature determine what exists and what cannot exist? Like you, the scientist must look for order in what appears complex and even chaotic. You can be confident that nothing happens in nature without a reason and a result. At times, however, it takes a pretty skillful observer to make sense out of what appears to be nonsense. Look for order, search for a reason, and expose the secrets of freshwater life.

A Tip on Procedure: How to Begin At this point the question may occur to you, "How does one discover these 'laws of nature'?" If this is your question, you have already found the answer, because *the answer is in the questioning.* Let us try to explain this paradox.

The secrets of nature are sometimes revealed through chance observation by someone who happens to be in the right place at the right time. This is the exception, however, not the rule. Scientists normally set goals in the form of questions. They seek to reach the goals by finding answers to the questions. Therefore, a scientist never looks at something to be studied without at least a few relevant questions in mind.

Let's look at an example. A scientist on board a research ship in the Arctic finds an unusual eel-like creature in a net. As he places the eel on the ship's deck, questions automatically flash through his mind. One very basic question (applicable, in fact, to any living thing) is, "How is the organism equipped for obtaining food?" This question has many aspects. It could read, "How does the organism locate, capture, dispatch, chew or swallow, and digest food items?" If the eel has no eyes, the question "How does it find its food?" may turn into a real problem.

Returning to the paradox, the answer to the question "How do we discover the laws of nature?" lies, initially at least, in asking the right questions. Once you start asking the right questions, you have a chance of finding the right answers.

For Thought and Research

1 Explain, in your own words, the definition of pollution that appears in this section.
2 List and explain five ways in which man's actions are governed by his environment.
3 Man attempts to control his environment through such processes as cloud seeding

(to produce rain) and crop spraying. List five additional methods by which man attempts to control his environment. Discuss the wisdom or folly of each method.

4 What do you think the "laws of nature" might be?

Recommended Readings

1 *Man in the Web of Life* by J. H. Storer, New American Library, 1968. This small pocketbook is easy to read and should increase your knowledge of man in relation to his environment.

2 *Concepts of Ecology* by E. J. Kormondy, Prentice-Hall, 1969. Read the section on environmental pollution in this book.

1.2   THE ECOSYSTEM CONCEPT

The word *ecology* is derived from two Greek words which, when put together, mean "a study of the home." When you study ecology you are learning about living organisms in their "homes" or *environments*. Today ecologists are becoming more and more important in a world which desperately needs conservation, reforestation, land improvement, flood control and, in general, more knowledge about ecological relationships. Lacking this knowledge, many people have in the past acted impulsively, and, even where they had good intentions, succeeded only in disturbing the balance of nature. For example, a farmer, hearing a noise from his hen-house one evening, finds a fox running off with a young chicken. He tells himself that from now on he must trap or shoot every fox he sees. In the long run, he is only making trouble for himself. For every bird which a fox may kill, it will also devour hundreds of mice, shrews, and rabbits, which are pests of the farmer in other ways. If the foxes in the area are wiped out, there could be an immense increase in the numbers of rodents and rabbits. This, in turn, would result in depletion of the grain crop and increased damage to stored grain. By shooting foxes, the farmer is destroying a vital link in an ecological chain and the system loses its balance.

We live in a world which is made up of ecological systems or ecosystems. An *ecosystem* consists of groups of organisms together with their non-living or physical environment. It is an interacting system. The living organisms in an ecosystem—plants, animals, and protists—are referred to as the *biotic* components. The non-living portion, or *abiotic* components, include soil, water, air, and physical factors such as light, wind, and temperature. Think for a moment about the various parts of the ecosystem in which we have an integral role. We rely on plants both directly and indirectly. Not only are they eaten by us, but they also are food for animals. In turn, the animals provide us with portions of our daily meals. The physical factors, like temperature, precipitation, and soil, determine the quantity and the quality of plants available, thus controlling our existence.

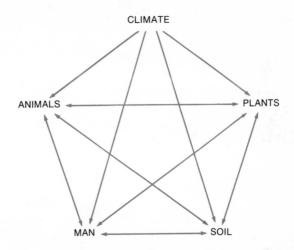

Fig. 1.2
The ecosystem concept.

Figure 1.2 illustrates an important aspect of an ecosystem. Consider each component individually and ask yourself how it is affected by any of the components which have arrows pointing to it. Then discuss your conclusions with others.

The most important thing about an ecosystem is that all members are closely interdependent. Now you can begin to see the importance of each individual's role. An ecological role is called a *niche*. Each niche is occupied by a different type of plant or animal. For example, it is the role of the green plant to produce food for itself and, in turn, for some animals. Thus in a natural community, a green plant occupies the niche of *producer*. Different plants occupy the same niche in different ecosystems. For example, a species of alga may control the life in a lake, while maple trees occupy the same niche in a forest.

Producers are the first and the most important link in the ecosystem. They provide food for the members which occupy the niches of *consumers*. There are two major types of consumers— *herbivores*, those which feed directly on green plants, and *carnivores*, those which feed on other animals. Let's look at the members of a grassland ecosystem to distinguish the producers and the consumers. Grasses are *producers*. You can probably name 20 or more herbivores which commonly feed on grass. Grasshoppers, groundhogs, deer, cattle, and rabbits are just a few of these. Since

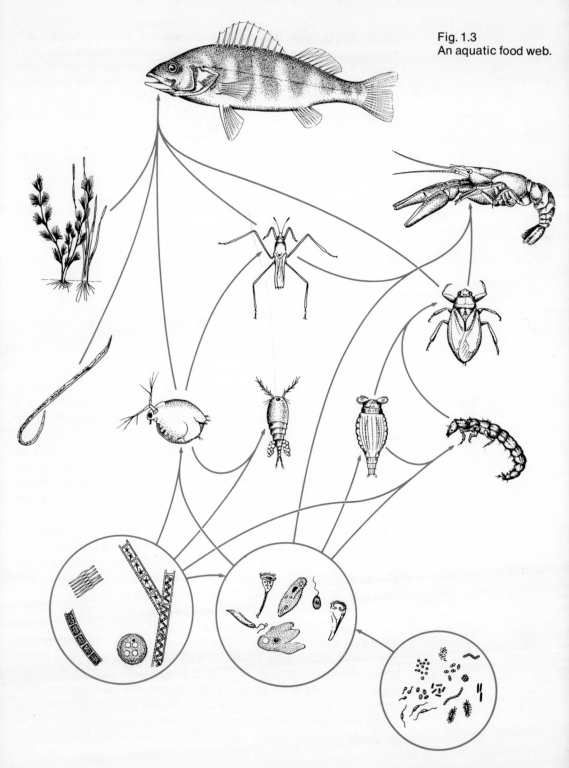

Fig. 1.3
An aquatic food web.

these animals feed directly on the producers, they are termed *primary consumers* and occupy the niche of *herbivores*. Animals such as frogs, preying mantises, and some birds prey on grasshoppers. These *first-order carnivores* which feed directly on the herbivores are called *secondary consumers*. See if you can name secondary consumers for each of the other herbivores mentioned above.

One of the most common predators of the frog is the snake, which is a *second-order carnivore*, or *tertiary consumer*. In many ecosystems there may be several levels of carnivores, but an essential member of each chain is the *top carnivore*, which has no predators. Can you name some top carnivores?

The life forms in an ecosystem are all linked together through predator-prey relationships in what is called a *food chain*. All food chains follow this general pattern:

$$Producers \rightarrow Herbivores \rightarrow Carnivores$$
$$\rightarrow Higher \ Order \ Carnivores$$

Within one ecosystem there may be over a hundred food chains. A *food web* integrates all of these individual chains into one diagram which may have a very intricate and complex pattern (Fig. 1.3).

As you think about food chains, you can see that most predators are limited to prey of a rather narrow size range. For example, a hawk must prey on an organism the size of a squirrel, mouse, or small rabbit and not on a deer or a moose! Food chains tend to proceed from very small organisms to progressively larger ones. Also, the number of individual organisms tends to diminish as you move further along a food chain. For example, many thousands of ants are needed to maintain only one anteater. A *pyramid of numbers* (Fig. 1.4) is often used to represent this fact. Each step in the pyramid is called a *trophic level*, and each successive trophic level contains fewer and fewer individuals. However, although this type of pyramid dramatizes the fact that a large number of organisms is needed to support one organism in the next highest trophic level, this is not of fundamental importance. This is because the pyramid of numbers treats individuals of all species as equal units, whether they are amoebas, cabbage plants, or cows.

Perhaps the construction of a *pyramid of biomass* (Fig. 1.5) makes more sense. Here the total mass of various organisms per unit area is considered. This pyramid also has one major fault—it equates unit masses of all organisms. For example, it implies that one gram of rabbit provides the same energy to a consumer as does one gram of grass. Yet experiments have shown that this is not true. Different organisms yield quite different amounts of energy. The copepod *Calanus* has a caloric equivalent of 7.4

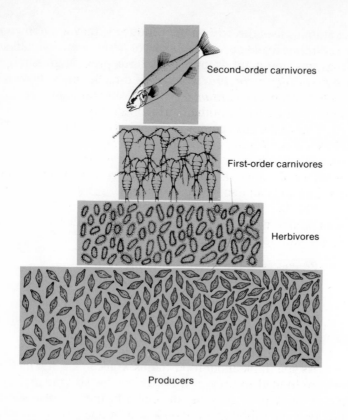

Fig. 1.4
A pyramid of numbers.

Second-order carnivores

First-order carnivores

Herbivores

Producers

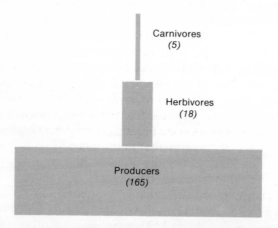

Carnivores
(5)

Herbivores
(18)

Producers
(165)

Fig. 1.5
A pyramid of biomass for a lake. The figures are the number of grams of biomass per square meter.

kcal per gram whereas the mollusc *Ensis* has a caloric equivalent of only 3.5 kcal per gram. Thus, in order to consume an equivalent amount of energy, a fish would have to eat at least twice the mass of *Ensis* as it would of *Calanus*. (A caloric equivalent of 1 kcal per gram means that complete burning of 1 gram of the substance releases 1 kcal of heat energy.)

Apparently we have only one choice left, the construction of a *pyramid of energy* (Fig. 1.6). This pyramid shows the energy

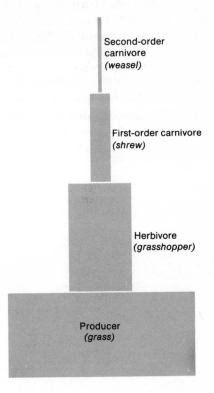

Fig. 1.6
A pyramid of energy. The width of each rectangle represents the total energy flow at that trophic level.

flow from one member of a food chain to the next. It also shows the role which each member plays in the transfer of energy. As you proceed upward on the pyramid, the members in each successive level conserve less energy to pass on to the next member. The weasel at the top of the pyramid in Figure 1.6 only conserves about 5% of all energy passed on to him from his prey. Any carnivore consuming only weasels would have difficulty taking in enough energy to exist. Why?

Two facts about energy are needed to understand the functioning of an ecosystem. First, there is a *one-way flow of energy* through the system. Energy must enter every ecosystem from the outside (from sunlight). Second, as you have seen in the pyramid of energy, a large amount of *energy is lost* from each level. Only about 15% is lost by plants, but about 95% is lost by weasels. It is used up in transpiration, growth, respiration, and other life processes. Every living organism requires energy in order to exist.

Although an ecosystem cannot function without the input of energy, the input of energy alone cannot make an ecosystem function. Over 20 different chemical elements must also be present before life processes in an ecosystem can be sustained. The chief elements required are carbon, hydrogen, oxygen, nitrogen, phosphorus, and sulfur. What roles do each of these elements play in the life processes of plants and animals? (If you have not yet studied enough biology to permit you to answer this question, your teacher will recommend a biology book that will help you to determine the answer.)

Unlike energy flow, the flow of these elements is cyclic within an ecosystem. This means that a constant input of the ele-

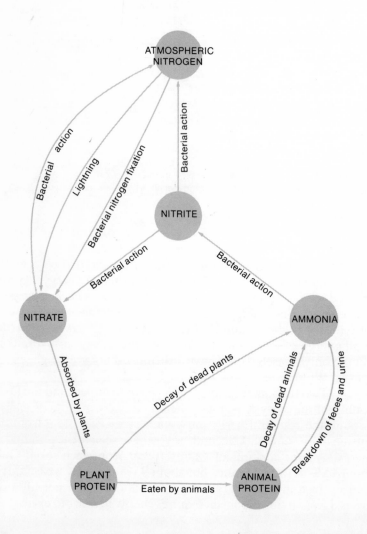

Fig. 1.7
The nitrogen cycle.

ments is not needed. Producers pass the elements on to herbivores, and herbivores pass them on to first-order carnivores. Thus the elements pass along the food chain. How do they get back to the producers to complete the cycle? An important group of organisms occupying the niche of *decomposers* performs this task. Decomposers are largely bacteria, yeasts, and fungi such as molds. They decompose, or break down, dead organisms and their waste products into simpler substances. These can be taken in and reused by producers (green plants). Frequently a group of microorganisms called *transformers* must act on the materials formed by the decomposers before these materials can be used by plants.

The nitrogen cycle (Fig. 1.7) illustrates the roles that decomposers and transformers play in an ecosystem. Ecologists call such a cycle a *nutrient cycle* or *biogeochemical cycle*. If you break down the latter term (*bio-geo-chemical*) and keep it in mind as you study the nitrogen cycle in Figure 1.7, you should be able to figure out why ecologists chose that term.

Although they play a minor role in the ecosystem, parasites, scavengers, and saprophytes deserve mention. *Parasites*, which may be carnivorous or herbivorous, live off their plant or animal hosts without actually killing them. *Scavengers* such as turtles, vultures, and crows feed on dead matter, breaking it down to a point where decomposers and transformers take over. Finally, *saprophytic plants* are a special group of scavengers. This niche is filled largely by fungi, such as mushrooms, which obtain their energy by absorbing organic matter from the soil, a tree stump, a decaying log, or some other dead object.

In this long section, you have met many new terms and concepts. All of them are needed to understand the workings of an ecosystem. The three major points to remember are these:

1) In an ecosystem, each member is an integral part, without which the system cannot function.

2) Energy flow is one-way. The energy passes from one level to the next in a food chain and a certain amount is lost at each level. (Follow the broken blue arrows in Figure 1.8.)

3) There is a continuous cycling of nutrients, which are essential to every individual in the ecosystem. (Follow the solid blue arrows in Figure 1.8.)

Now that you have read about ecosystems, set up one of your own, as described in the next section.

Fig. 1.8
Flow of energy and nutrients through an ecosystem. Solid arrows represent cycling of nutrients. Broken arrows represent energy flow.

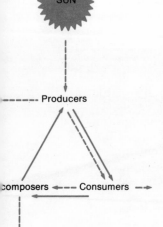

For Thought and Research

1 Complete these food chains:
   (a) In a meadow: *grass* → *grasshoppers* → . . .
   (b) In a forest: *tree leaves* → *plant lice* → . . .
   (c) In a pond: *microscopic algae* → *microscopic animals* → . . .
   Compare your results with those of others and draw appropriate conclusions.

2 You will find quite often that an organism cannot be placed exclusively into any one niche, but may occupy two or even three at the same time. Consider, for example, the Venus fly trap, a green plant which manufactures food of its own and yet is a predator of flies and other small insects. Another example is the striped skunk, which normally eats a meal of grubs, insects, worms, berries, tender roots of plants, and your garbage. Name the niches which these two organisms occupy. Can you think of other examples? Now consider man. Is there any ecosystem of which he is not a part? What roles does he play? List five food chains in which man is the last link.

3 *Phytoplankton* (plant plankton) are microscopic plants that are the main producers of most aquatic ecosystems. *Zooplankton* (animal plankton) are minute animals that, in most cases, feed on phytoplankton.

   What do you conclude from the pyramid of biomass shown in Figure 1.9? Of what significance is the fact that it is inverted? (The first trophic level is normally the largest in a pyramid of biomass.)

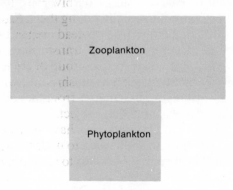

Zooplankton

Phytoplankton

Fig. 1.9

4 Summarize the carbon cycle from one of the *Recommended Readings*. Pay particular attention to the roles of producers, consumers, and decomposers in this cycle.

5 (a) Why do pyramids of energy always become more narrow as one moves to higher and higher trophic levels?

   (b) In this section you were asked why a carnivore that consumed only weasels would have difficulty taking in enough energy to exist. If you couldn't answer the question, try again, keeping in mind your answer to (a) and the pyramid of energy shown in Figure 1.6.

6 Perform the following laboratory studies to increase your knowledge of the basic ecological principles presented in this section:
   5.22 *Effects of Crowding on a Population*
   5.27 *Laboratory Study: Analyzing Gut Contents*
   5.29 *An Experimental Food Chain for the Laboratory*

Recommended Readings

The three books which follow are simple, introductory books on ecology. They deal with the basic principles that are introduced in this section. You will find any one of them enjoyable and worthwhile reading.

1   *Basic Ecology* by Ralph & Mildred Buchsbaum, Boxwood Press, 1957.
2   *Ecology* ed. by Peter Farb, Life Nature Library, Time Inc., 1963.
3   *The Web of Life* by J. H. Storer, New American Library, 1953.

The following books are more difficult but also more informative. You may want to look up specific things like pyramids of biomass, energy flow, and nutrient cycling in one or more of these books.

4   *Readings in Conservation Ecology* ed. by G. W. Cox, Appleton-Century-Crofts, 1969.
5   *Ecology* by E. P. Odum, Holt, Rinehart & Winston, 1963.
6   *Elements of Ecology* by G. L. Clarke, John Wiley & Sons, 1966.
7   *Concepts of Ecology* by E. J. Kormondy, Prentice-Hall, 1969.
8   *Fundamentals of Ecology* by E. P. Odum, W. B. Saunders Co., 1959.
9   *Ecology and Field Biology* by R. L. Smith, Harper & Row, 1966.

1.3   A CLASSROOM ECOSYSTEM

For this experiment, you need a large jar or bottle which can be stoppered and made airtight. Its capacity should be at least 2 or 3 gallons. Place a small amount of sand in the bottom and fill the jar with water. If you use tap water, let the water dechlorinate for 48 hours or more. Add a few strands of an aquatic plant such as *Elodea* and, if available, a few spoonfuls of a small floating aquatic plant such as duckweed. Now place a few snails and one or two small fish (such as guppies) in the water. Stopper the bottle and place one or two 100-watt light bulbs in the position shown in Figure 1.10.

Observe your ecosystem for almost a full year. During the first few weeks, you may have to change the positions of the lights and add or remove plants and animals until a balance is attained. One factor should be kept in mind when making your observations. Your ecosystem is a *closed* ecosystem. How does this differ from naturally occurring ecosystems? What advantages are to be gained by studying a closed ecosystem? What niche does each of the living organisms, plant and animal, occupy in this system? What nutrient cycles are taking place? What niches are occupied by microorganisms that you cannot see? Account for any changes that occur in the ecosystem as time progresses. What evidence do you have that energy flow is unidirectional in ecosystems?

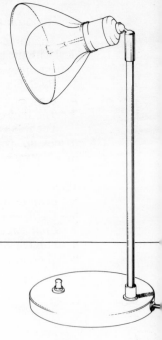

Fig. 1.10
A classroom ecosystem.

For Thought and Research

Make preparations to begin the studies outlined in Section 5.19—*A Study of a Mini-ecosystem* and Section 5.20—*A Study of a Mini-ecosystem (Advanced)*. Begin these studies as soon as you have read enough of this guide to understand what you are doing and why.

Limnology: The Nature of Ponds and Lakes

2

All of the evidence indicates that life originated in water. And it is an established fact that all life must have water to survive. From the land spider, to the trout, to the whale, all life requires fresh water. (How do you think the whale acquires it, amid all that salt water?) Thus, water is man's most important asset. It affects his cultural and his social life, his industry, his agriculture, his leisure, and his war.

Our main interest, however, is a study of water from an ecological standpoint, that is, of water and of the life in and around it. Imagine yourself standing on the shore of a lake or pond. What forces are at work within this ecosystem? How do the forces act? Why do they operate in these ways? For that matter, what is water, anyway?

2.1   WATER—A TOPSY-TURVY COMPOUND

Study a glass of water. What are its characteristics? It is a liquid at room temperature, it is tasteless, odorless, and colorless. These facts and many others have been familiar to you for years. Now look again. Notice anything unusual? No? Place the glass of water

18

in a freezer, and examine it in an hour. Now do you notice anything unusual? A thin layer of solid water (ice) is at the top. That in itself does not seem odd. But, if some carbon tetrachloride were to be frozen (−22.8°C) you would not find solid forming at the top. *It would start forming at the bottom*. This difference between water and most other liquids is probably basic to our existence. Most other liquids are heaviest (densest) at their freezing points, and sink as they freeze. Water, however, is densest a few degrees above its freezing point (about 4°C). This dense water sinks and the lighter frozen water (ice) floats. Aquatic life can thus continue under the ice. If water froze like most other liquids, the aquatic life in most bodies of water would be slowly lifted to a death at the surface. The ice also insulates the water below it, preventing it from freezing. If ponds and lakes were to freeze from the bottom upwards, most of them would be solid ice during the winter, since no insulation would be formed to prevent this.

For Thought and Research

1 Why is the solid state of most substances *more* dense than the liquid state?
2 Why is water *less* dense in the solid state than it is in the liquid state?

Recommended Readings

The following books contain information on the topics that you are asked to investigate in the *For Thought and Research* section.
1 *Interaction of Matter and Energy* by N. Abraham et al., Rand McNally, 1969.
2 *Modern Chemistry* by H. C. Metcalfe et al., Holt, Rinehart & Winston, 1970.
3 *Chemistry: Experimental Foundations* by R. W. Parry et al., Prentice-Hall, 1970.

2.2 TEMPERATURE AND DISSOLVED OXYGEN— FOLLOWING THE LEADER IN A WATERY WORLD

One of the major factors acting on a pond or lake is the climate. Since the climate changes according to the season, it produces seasonal changes in lakes and ponds. These changes are of fundamental importance to us.

To understand these important changes, it may be best to obtain a fish's-eye view. Imagine, then, that you are a fish! It is early spring and you are a recently hatched trout. The water is cold and, therefore, comfortable. In addition, there is quite a bit of dissolved oxygen in the water. Breathing is easy.

As spring progresses, you begin to explore the world. It is about 100 feet deep. The temperature is fairly constant throughout, except, perhaps, in the lower 10 feet or so, where it is a little warmer (about 1 or 2 C°) than elsewhere. The dissolved oxygen content is also fairly constant throughout (Fig. 2.1). Very often, in the spring, the water appears to be in constant motion throughout its depth and breadth. Materials are seen to be "picked up" from the bottom and carried upward to be dispersed. This picques your curiosity so that you consult your school leader in order to get a better understanding of your world. And this is what he tells you:

"Since we live in water, it is very important to understand its phenomena. As you have seen, water temperature and dissolved oxygen are fairly uniform, or homogeneous, from top to bottom in the spring season. In addition, water density is also homogeneous, since it depends on temperature. Therefore, it is possible for the winds above our world to thoroughly mix the water. This distributes materials and nutrients equally throughout. The diagram that I have made in the sand shows how this happens. [Fig. 2.2.] Atmospheric gases, including oxygen, are mixed in a like manner. The nutrients and the oxygen stimulate and support the growth of young plants. During their complicated food-producing process of *photosynthesis*, these plants produce oxygen. This, too, is mixed uniformly thoughout the water.

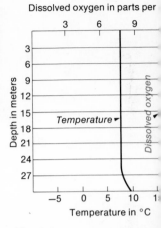

Fig. 2.1
Graph depicting temperature and dissolved oxygen conditions in a lake in spring and fall.

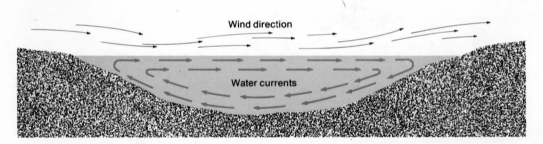

Fig. 2.2
Diagram showing how winds cause water currents which can thoroughly mix the water during spring and fall.

"These conditions are true for our world and for others that are a little deeper or a little shallower. But, they are not true for very deep lakes. There only the upper 100 feet or so of the lake undergo this process known as the *spring overturn*. Have you followed me so far?"

"Yes, I think so," you answer, "but will our world always be like this?"

"No, it won't," answers your school leader. "As summer approaches, the sun warms the upper layer of water faster than the wind can mix it. This warmer water, being less dense than the colder water below it, tends to remain 'floating' on the surface. Thus two layers develop as you can see in this diagram [Fig.

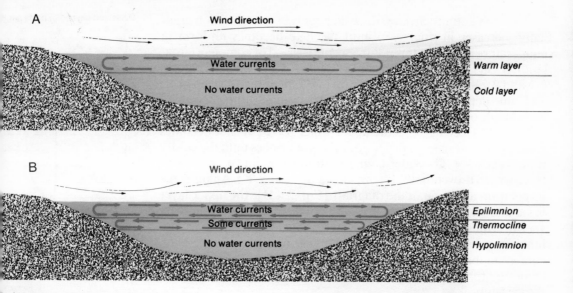

A

Wind direction

Water currents

No water currents

Warm layer

Cold layer

B

Wind direction

Water currents

Some currents

No water currents

Epilimnion

Thermocline

Hypolimnion

Fig. 2.3
Diagrams illustrating the stratification occurring in a lake (A) in late spring and early summer, and (B) in mid summer.

Fig. 2.4A
Temperature and dissolved oxygen conditions in a lake in summer.

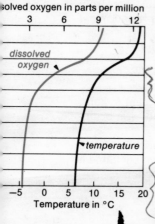

Dissolved oxygen in parts per million

3   6   9   12

dissolved oxygen

temperature

−5   0   5   10   15   20
Temperature in °C

2.3 A]—a warmer, upper layer, and a cooler, lower layer. Eventually three distinct layers form in this process of *thermal stratification*: an upper, warmer, freely-circulating *epilimnion*; a layer of transition from warmer to colder waters called the *metalimnion* or *thermocline* (defined by a drop in temperature of 1.0C° for every depth increase of 1.0 meters); and the lower, colder *hypolimnion* of relatively uniform temperature. This layer is cut off from circulation and hence receives no atmospheric gases. Now, can you point out these three layers in the diagram in the sand? [Fig. 2.3 B.]

"Very good! Now, these layers occur in waters which are quite deep (over 40 feet). In shallow lakes and ponds the water is usually warmed uniformly and completely." Pointing to yet another diagram (Fig. 2.4A), your leader explains: "This diagram shows the temperature and dissolved oxygen (D.O.) content of our lake in summer. The D.O. level has a curve similar to that of temperature. This is true for most lakes, but there are exceptions. In some cases, the thermocline will have the greatest D.O. content. Can you figure out why this should be so?"

"Well, I'm not sure," you answer. "Does it have something to do with how much gas can dissolve in water at different temperatures?"

"That's good! The colder the water is, the more gas it can dissolve. Furthermore, organisms needing oxygen for life are usually more abundant in the epilimnion. Therefore, they tend to reduce the D.O. content there.

"As autumn arrives, the water cools. The epilimnion gradually increases in thickness, until the lake becomes uniform in temperature again. At this point a second period of complete circulation occurs that is called the *fall overturn*. [Look back to Figures 2.1 and 2.2.] This usually lasts from late September to December. The time depends on the area of the lake, and on the geography and climate of the region.

"In December, as the upper water layer cools, it becomes more dense and sinks to the bottom. This continues until the colder water reaches 4°C. Water colder than this remains near the surface, where it eventually freezes. [See Section 2.1.] Again three layers form as shown in the diagram [Fig. 2.4 B]: the upper layer is ice and water close to the freezing point; the thermocline is reversed, going from colder to warmer; the lower layer (hypolimnion) is, once more, of uniform temperature. Usually the D.O. content in the hypolimnion drops off sharply in mid-winter. It becomes uniform once more during the early spring overturn after the ice has broken.

"Most waters go through this cycle each year. In areas with warm climates year round, the winter stratification (layering) may not occur and the lake undergoes only a summer stratification and an overturn. Now, do you understand this? Do you see why you must avoid the epilimnion in the summer and why you shouldn't enter the lower part of the hypolimnion?"

"I guess so," you answer.

"Don't guess! Know! Your survival depends on it!"

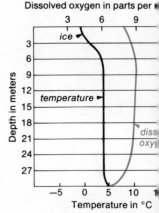

Fig. 2.4B
Temperature and dissolved oxygen conditions in a lake in winter.

For Thought and Research

1 In many waters, trout and other fish migrate seasonally. They move into deeper water in the summer and into shallower water in the winter. Why is this so?

2 If a small lake becomes thermally polluted through the addition of hot water from a power generating plant, it may remain in a "summer condition" all year round. What are the likely effects that this will have on the lake ecosystem?

3 Perform the laboratory study outlined in Section 5.17—*A Model of a Lake*.

Recommended Readings

1 *A New Field Book of Freshwater Life* by Elsie B. Klots, G. P. Putnam & Sons, 1966. The first 25 pages deal with the characteristics of lakes and ponds at a beginner's level.

2 *A Treatise on Limnology* by G. E. Hutchinson, John Wiley & Sons, 1957. One of the most cited books in studies of fresh waters. It is quite advanced but you will find it useful for extending your knowledge of the concepts introduced in this section.

3 *Ecology and Field Biology* by R. L. Smith, Harper & Row, 1966. This book contains a thorough and interesting chapter on lakes and ponds.

2.3 PHOTOSYNTHESIS AND RESPIRATION— MAINTAINING A LIFE BALANCE

As was suggested earlier, aquatic plants play a role in the oxygen supply of water.

The dissolved oxygen content of water depends on several factors, including temperature, light, living and dead organisms, and other organic matter present in the water. Oxygen enters water in two ways: directly from the atmosphere (see Section 2.2), and from plant photosynthesis. It is removed primarily by the respiration of producers, consumers, and decomposers.

Plant photosynthesis is a complicated process whereby plants make organic compounds, such as glucose, from carbon dioxide, water, and light energy. Oxygen is released as a by-product. The process can be carried out only during periods of light. Even then the light must be of the right wavelengths and intensity. If, however, all of the conditions are met, oxygen is given off, and is dissolved in the water.

During respiration, animals and plants combine foods like glucose with oxygen, and give off carbon dioxide and water as waste products. At the same time, energy is released to enable life processes to continue. Oxygen is used likewise by many decomposer organisms as they break down dead animals and plants (Fig. 2.5). These processes of respiration occur at all depths in lakes and

Fig. 2.5
The diagram illustrates how plants, animals, and decomposers maintain each other: the plants by giving off oxygen (solid arrows) which is used by fish and decomposers for their life processes; the fish and decomposers by giving off carbon dioxide (broken arrows) which is used by the plants.

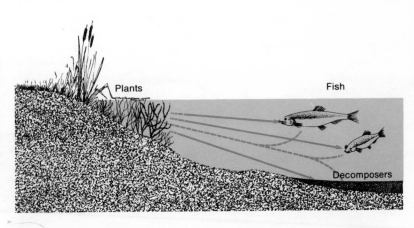

Plants

Fish

Decomposers

ponds. Therefore, carbon dioxide is produced and oxygen is consumed at all depths. However, since most dead and waste materials sink to the bottom, most of the decomposition occurs in the lower part of the hypolimnion. This is called the *tropholytic zone.*

(*Troph-* means "nutrient" and *-lytic* means "decomposition." Organisms in this zone decompose nutrient matter.) Thus there is a gradual decrease in dissolved oxygen content as you go deeper into the hypolimnion. In some lakes no dissolved oxygen is found in the tropholytic zone. Decomposition occurs there through *anaerobic* means (requiring no oxygen). This results in the formation of carbon dioxide, methane, and hydrogen sulfide ("rotten egg") gases. Hydrogen sulfide can be found easily by pushing a stick into the bottom ooze near the shore of a very fertile lake, and then pulling the stick out.

Although carbon dioxide is given off at all depths, oxygen is not. Remember that oxygen is formed during photosynthesis and that light is required. There is a limit to the depth to which light can penetrate water. Depending on the conditions of the water—its murkiness—light penetration differs for each pond and lake. Thus oxygen is *not* given off at all depths, but only in the upper layer— the *trophogenic zone*. (*Troph-* means "nutrient" and *-genic* means "producing." This zone is characterized by the production of nutrients by green plants.) The depth of this zone may vary from 5 to 90 feet and, in many cases, corresponds roughly to the epilimnion. (Do you remember what that is?) Between the trophogenic and tropholytic zones lies a layer called the *compensation depth* where oxygen production is balanced by carbon dioxide production.

To further complicate this picture, ecologists divide the trophogenic zone into two subzones. The *littoral zone* (*littoral* means "pertaining to the shore") is defined as that region where rooted plants grow and hence where light penetrates to the bottom. The *limnetic* or *open-water zone* (*limn-* means "lake") is that which extends to the depth of effective light penetration (Fig. 2.6).

Fig. 2.6
The major zones of a typical lake. In the littoral zone (rooted plants) and the limnetic zone (phytoplankton) oxygen production is greater than carbon dioxide production. Note that these two make up the trophogenic zone. The compensation depth is a layer of transition from the trophogenic zone to the tropholytic zone where carbon dioxide production is greater than oxygen production.

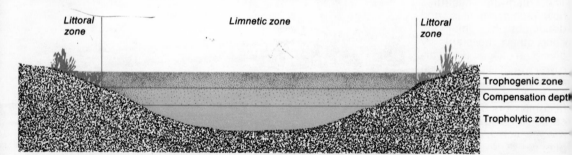

| Littoral zone | Limnetic zone | Littoral zone |

Trophogenic zone
Compensation depth
Tropholytic zone

Although the littoral zone is more interesting as a study area, the limnetic zone is more important in terms of oxygen production. It is usually the larger in area and most *phytoplankton* (tiny plants of microscopic size) are found there. These plants are the lake's principal oxygen producers.

For Thought and Research

1 Investigate further the processes of photosynthesis and respiration by consulting one or more of the *Recommended Readings*. Write the summation equation for each process and account for each term in the equation. What relationship exists between the two equations?
2 What would be the immediate effect of an algal bloom in a lake? What would be the long-term effect? Explain your answers.
3 Substances like copper sulfate are often added to nutrient-rich lakes to control algal blooms. Discuss the pros and cons of this procedure.
4 Why is it important to know where the littoral zone of a lake or pond is? What animals (including fish) and plants would you expect to find in the littoral zone?

Recommended Readings

For further information on photosynthesis and respiration consult:

1 *Biological Science: Molecules to Man,* B.S.C.S. Blue Version, Houghton Mifflin, 1969.
2 *Biological Science: An Inquiry into Life*, B.S.C.S. Yellow Version, Harcourt Brace Jovanovich, 1969.
3 *High School Biology*, B.S.C.S. Green Version, Rand McNally, 1969.
4 *The Spectrum of Life* by H. A. Moore and John R. Carlock, Harper & Row, 1970.
5 *Modern Biology* by J. H. Otto and A. Towle, Holt, Rinehart & Winston, 1968.

The references at the end of Section 1.2 should be consulted for more information on zonation in ponds and lakes.

2.4 OTHER FACTORS

The factors at work within a body of water are complex, but can be decoded. A pattern can be found for all of the factors so far discussed. Other factors are not so easily intelligible, yet even so, a pattern of sorts can be observed if you look for *relationships*. Do not look at any one factor in isolation from others of equal importance.

All of these factors are *limiting*, that is, each has a maximum and a minimum value, above or below which life, for many species, cannot continue. Those species that do survive beyond these limits are generally of decreased commercial, recreational, and aesthetic value.

The *p*H of water is one such factor. It expresses the concentration of hydrogen ions in a solution on a scale that runs from 0 to 14. On this scale 7 is neutral, below 7 is acidic, and above 7 is basic. In other words, a solution becomes less acidic and more basic as the *p*H goes from 0 to 14. The chemistry behind the value of pH does not concern us here. You can, however, read about it in almost any high school chemistry text.

What *does* concern us here is the limiting nature of pH. Although many fish species can tolerate a pH range from 5 to 9, most have a much narrower range. Indeed, some cannot survive a change of even 1 pH unit. Only a few species can tolerate pH values over 9 or under 5 and these are usually specifically adapted to these extreme conditions. Such organisms commonly lose the ability to tolerate pH values between 5 and 9.

It is an interesting fact that the pH of a body of water generally changes as the body of water ages (see Section 3.4). Usually basic when it is young, it becomes more acidic as it gets older. This is primarily due to a buildup of organic materials which, as they decompose, release carbon dioxide gas. This gas, in turn, reacts with the water to form hydrogen ions, the cause of acidity. In response to these changes, the life within the pond or lake will also change as aging proceeds.

Closely allied with pH and in many ways determining it are the *alkalinity* and the *hardness* of the water. Alkalinity is a factor that is dependent upon all of the bases (alkalies) and basic salts that are present in the water. Hardness is caused almost entirely by the salts of calcium and magnesium that are dissolved in the water. Again, we recommend that you read further about these factors in chemistry books. Our prime concerns here are the sources of these chemicals and their effects on living organisms.

Most lakes and ponds receive water from various sources—runoff from rain, direct rainfall, streams and rivers, and underground springs. These waters *leach* out (break off and then dissolve or carry as a suspension) many chemicals from rocks and soil, and carry them into bodies of water. Some of these chemicals then enter into their own particular biogeochemical cycle (see Section 1.3). Also, falling rain drops pick up many chemicals from the air and carry them into lakes and ponds (Fig. 2.7). These two processes also affect other factors, which are closely related to each other and to the factors already mentioned. These other factors will be discussed next.

When you look at the water in a lake or pond, what do you see? Is the water brown, clear, colorless, murky, silty? The answer is, of course, that at different times the water can have any of these appearances. For example, after a heavy rain the water appears murky or just plain dirty. This results from material carried into it by the surface runoff. Methods have been developed for measuring the amount of material present in the water. In general, the results obtained are fairly constant for a particular pond or lake. These tests are not done just after a rain storm, since the abnormal condi-

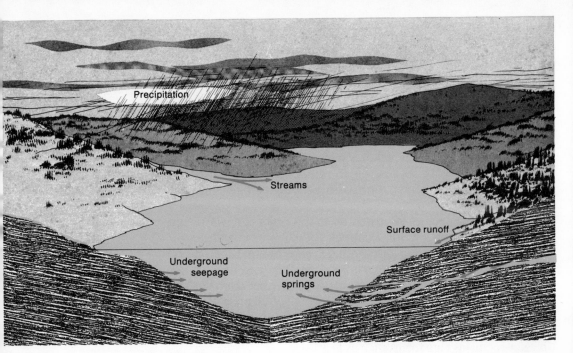

Fig. 2.7
Diagram illustrating how water carries chemicals into ponds and lakes.

Labels on diagram:
Precipitation
Streams
Surface runoff
Underground seepage
Underground springs

tions brought on by the storm will usually remain for a day or two. The reason for these measurements is to determine the *productivity* of the lake or pond, that is, its ability to support life. Knowing the productivity, a biologist can calculate how much life the pond or lake can support (usually in pounds or tons of fish per unit volume). A lake with high productivity and fertility is said to be *eutrophic*. It has a greater abundance of materials than a lake with low productivity and fertility (an *oligotrophic* lake). To determine productivity, measurements are made of total dissolved solids (T.D.S.), conductivity, total suspended solids (T.S.S.), and turbidity.

Total dissolved solids (T.D.S.) is a measurement, by weight, of the total amount of material dissolved in a measured volume of water. These materials are usually present as ions in the water. They come out of solution as solids when the water is slowly evaporated. A great deal of research has been done recently on the relationship between total dissolved solids and another factor, the conductivity of the water. Since these materials exist as ions in the water, their solution can conduct an electrical current. The amount of *conductivity* is directly related to the concentration of ions and, hence, to the total dissolved solids. In the case of very high total dissolved solid values, however, the correlation does not seem to hold up. Also, the water may contain some dissolved solids that are not ionic and, as a result, do not contribute to the conductivity of the water.

The *total suspended solids* (T.S.S.), on the other hand, is a measurement, by weight, of the amount of material suspended in a measured volume of water. Suspended matter consists chiefly of phytoplankton, zooplankton (tiny animals), silt, human sewage, and a vast range of wastes from animals, plants, and industry.

Closely related to T.S.S. is a factor called *turbidity*. A turbidity reading measures the ability of a light beam to pass through the water sample. Water which has materials suspended in it scatters and absorbs light rays entering it, rather than transmitting the light in straight lines. Thus, a turbidity reading can give an indication of the T.S.S.

All of the factors discussed up to this point can drastically alter a lake or pond. If, for example, a lake is too fertile, *algal blooms*, the sudden appearance of huge crops of algae, can occur after the spring overturn. They choke the water, and make it not only unsightly, but hazardous to man, fish, and other animals. The fact that algal blooms are dangerous to fish may surprise you. But when the algae die, an enormous amount of oxygen is required for their decomposition. This leaves little or no oxygen for *aerobic* (oxygen-requiring) life forms such as fish.

The last factor is one which we have already studied—*temperature*. In Section 2.2 we saw how temperature plays a major role in the dissolved oxygen content of water and, therefore, in the life content of a body of water. Temperature also exerts a direct effect on living organisms. Just as an organism can tolerate only a certain range in *p*H, it can also tolerate only a certain temperature range. This may vary, as in the case of some fish, through different stages of its life cycle. Further, different species have different temperature ranges. One species can tolerate a range from 35 °F to 65 °F, and another a range from 45 °F to 75 °F. Within the tolerated range, however, death may still occur if a sudden temperature change occurs. Fish may *acclimatize* (become accustomed) to higher or lower temperatures than their normal range if the change occurs slowly. A decrease in dissolved oxygen, an increase in carbon dioxide or the presence of toxic materials may greatly reduce the upper *lethal temperature* (the temperature at which death occurs) of a species. Why? Even when a change in temperature occurs which, by itself, should not be lethal, a species may be wiped out, because the new temperature may be more favorable to a competitor, a predator, a parasite, or a disease.

All of these forces, in combination, act to stratify (place in layers) most life forms in a pond or lake. Each species seeks that area which has the physical and chemical properties most favorable to it. For example, a fish which can live in a temperature range of 35 °F to 65 °F, may be found only in the 35 °F to 45 °F range, since its preferred *p*H and oxygen conditions are also found there. The next unit discusses the common life forms to be found in the various layers of ponds and lakes. Read it carefully and you will increase the "catch" of your field trips many times.

For Thought and Research

1 (a) Make a list of the factors that determine the *p*H of a lake that is not under man's influence.

(b) Make a second list of 5 or more ways by which man causes the *p*H of some lakes to change.

2 A certain species of trout is found in a lake that has a *p*H of 7.2. Another lake also has a *p*H of 7.2. Would you expect to find trout in it too? Explain your answer.

3 Find out (from a chemistry book or chemistry teacher) why the correlation between T.D.S. and conductivity breaks down for high T.D.S. values.

4 Why is the temperature requirement of a particular fish species different for each life stage (egg, fry, and adult)?

5 In what ways will the physical and chemical characteristics of a lake or pond differ from the normal after a heavy rainstorm?

6 Make a list of the names of 5 bases and 5 basic salts. What do all of these compounds have in common?

7 Why is water that contains salts of calcium and magnesium called "hard"? Distinguish between carbonate (temporary) hardness and permanent hardness.

Recommended Readings

1 *A Guide to the Study of Environmental Pollution* by W. A. Andrews et al., Prentice-Hall, 1972. This book contains further information on *p*H, alkalinity, hardness, temperature, D.O., carbon dioxide, and other physical and chemical factors.

2 *Fundamentals of Limnology* by F. Ruttner, University of Toronto Press, 1963.

3 *A Treatise on Limnology* by G. E. Hutchinson, John Wiley & Sons, 1957.

Titles 2 and 3 contain advanced discussions of the physical and chemical factors introduced here.

4 *Streams, Lakes, Ponds* by R. E. Coker, Harper & Row, 1968. This easy-to-read book discusses most of the physical and chemical factors introduced in Sections 2.2, 2.3, and 2.4.

Life in Lakes and Ponds

3

What can you expect to find and where? Although this is a very practical question, it is impossible to answer precisely. Some organisms will be present in any natural body of water. Bacteria and protozoa are the best examples. Other things will most likely be present, including certain types of plankton, aquatic plants, bottom scavengers, and insect predators; but which varieties is another question.

First of all, each lake and pond is unlike any other. Not only are the physical and chemical factors different, but so are the particular plant and animal communities. Each is a separate ecosystem with its own peculiarities. Your own skill at finding things is another important factor. Just because a certain form of life is present doesn't mean that you will find it. Bacteria and protozoa are microscopic and require special collection and observation techniques. Larger organisms will flee if you are clumsy. You won't even see them go. You must be thorough and careful in your searching techniques.

A person knowledgeable in aquatic life will undoubtedly find more than the beginner. The experienced person, although not knowing for sure what lives in a given pond or lake, will expect to find certain animals and plants given certain conditions.

In this unit, you will meet some of the common life forms

Fig. 3.1
Gerris, the water strider.

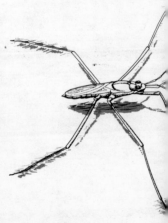

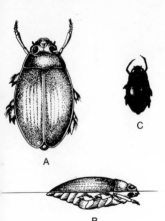

Fig. 3.2
Drawings of *Gyrinus* showing (A) an enlarged aerial view, (B) a side view showing the double set of eyes, and (C) the approximate natural size.

of the freshwater environment and learn where they are most commonly found. This is an attempt to answer the practical questions, "What?" and "Where?" Remember, however, that some organisms cannot be placed exclusively in one particular habitat. They do wander about. Also, each inland water area is different; what is true for most may not be true for all.

3.1　ON THE SURFACE

Most people are familiar with at least some of the animals that live on the surface of lakes and ponds. *Gerris*, the water strider (Fig. 3.1), is a familiar sight. It skates and jumps about on the surface film of ponds, quiet bays, and even slow sections of streams and rivers. Having six legs (not eight), it is an insect and not a spider with which it is sometimes confused. The front pair of legs point forward and down, and aid in feeding. All of the legs have waxy hairs on the jointed tips or "feet." The arrangement of these hairs and their waxy coating repel water and keep the insect afloat. The second and third sets of legs are more important in locomotion. Other common names for *Gerris* are pond skater and wherryman.

The whirligig beetle, *Gyrinus* (Fig. 3.2), is an unmistakable and talented quiet-water inhabitant. Gliding around and around like a little speedboat, it gives the impression that its life is a continuous holiday. An unusual feature of this beetle is its divided eyes. Half of each eye looks under water while the rest looks above. It watches both worlds at once as it glides about in search of insects that fall on or into the water. This is the adult, which most people see. It usually appears in late summer or autumn. Up until this time the larva has been actively feeding beneath the surface.

Less familiar surface dwellers include the springtail (Fig. 3.3) and the fisher spider (Fig. 3.4). The former is a tiny insect less than a quarter of an inch long. It is named after its most unusual characteristic. Its tail can be folded and set in place beneath its body like a safety pin. When the tail is released, it acts like a spring, catapulting the animal into the air. When you attempt to collect one of these tiny black specks from the water's surface, its spring goes into action and it vanishes, to reappear as much as a foot away. The fisher spider is sometimes seen in areas of dense aquatic vegetation, walking across water and vegetation with equal ease. When frightened it may disappear down a stem below the surface. Although perhaps the heaviest of the surface dwellers, its weight is well distributed over the surface film. The large number of body hairs tend to increase its own surface area. The body hairs, being water repellent, trap a thin jacket of air around the body when the

Fig. 3.3
With a downward thrust of its tail, the springtail can jump over a hundred times its own body length.

spider submerges. Breathing from this jacket, the spider may have enough oxygen to last half an hour under water.

The fisher spider is a casual visitor beneath the surface, where most aquatic life has taken up permanent residence. The next sections deal with organisms adapted to a wet way of life.

3.2 ON AND IN THE BOTTOM SEDIMENTS

Of all the inhabitants of the pond or lake, the bottom dwellers show the greatest diversity of form and life style. The reason is that a great variety of microhabitats can be found there, each one suited to particular modes of life. Another reason is the great quantity and variety of food present.

Things that die, both plant and animal, usually end up on the bottom. A whole host of organisms is involved in the consumption of this dead matter falling from above. Most important are the bacteria. They are often found in such great numbers that they use up all of the available oxygen for their own respiration. This makes it impossible for other organisms to survive in the bottom sediments. It is hard to detect bacteria in collected bottom material without using special culturing techniques. Thus, bacteria are often overlooked, even though their role in returning nutrients to the water helps to keep a lake or pond the active place that it is.

Various fungi, also important to organic breakdown, occur in the bottom detritus (material worn away from a mass). Even species of *Penicillium* can be found in aquatic habitats. Here again, the examination of these organisms usually requires special techniques.

Although protozoans (Fig. 3.5) occur in almost every drop of water in the lake or pond, most types are located on or near the

Fig. 3.4
The fisher spider.

Fig. 3.5
Some protozoans that commonly inhabit ponds.

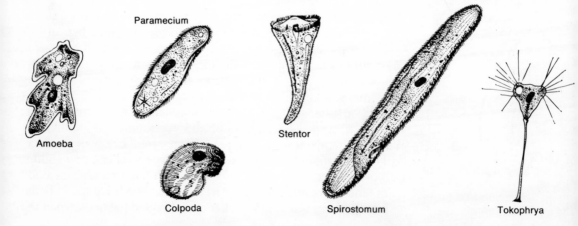

Paramecium

Stentor

Amoeba

Colpoda

Spirostomum

Tokophrya

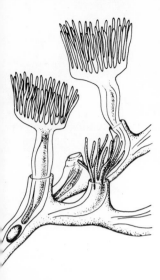

Fig. 3.6
A typical bryozoan.

'Foot'

Fig. 3.7
A mussel.

Fig. 3.8
Two mature *Hydra*, one "budding" new individuals from its body and the other digesting a water flea, are shown attached to a twig on the bottom of a pond.

bottom. They carry out activities similar to those of bacteria and fungi. The well known *Amoeba* is found here, slowly changing into all sorts of shapes and gobbling up organic material. The Suctoria, a group of protozoa that are attached to material in the pond, can also be found. This group feeds on other protozoa, killing them with poisonous tentacles through which they also suck the prey's protoplasm. *Tokophrya* is a representative of this group.

A number of other attached forms live in the bottom sediments or on detritus and bits of vegetation. Bryozoa (Fig. 3.6) and freshwater sponges deserve mention, although they are somewhat rare and easily overlooked in field studies. A typical sponge, for instance, may consist of little more than a very thin, slimy, green or brown coating on a twig. The bryozoans are colonial animals with a much more intricate structure than sponges. Their colonies may include hundreds of thousands of individuals and reach the size of basketballs. In either case, the most interesting aspects of sponges or bryozoans can be seen only under the microscope.

Another fairly immobile animal is the mussel or clam (Fig. 3.7). Although most common in large rivers, these animals do occur in wave swept sections of lakes where they filter tiny food particles from the water. They are capable of some movement using their single so-called "foot." Movement is most important with the approach of winter, when the animal must burrow deeper into the sediment to avoid the killing cold.

The most fascinating of the attached forms is the *Hydra* (Fig. 3.8). In Greek mythology, *Hydra* was the name of a monstrous water serpent having nine heads. When any head was cut

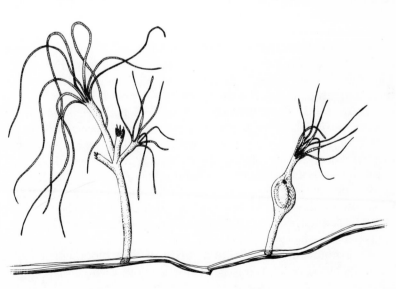

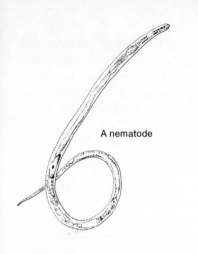

A nematode

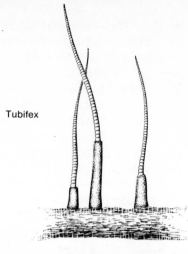

Tubifex

Fig. 3.9
Two types of worms that commonly inhabit the bottom sediment of a pond or lake.

off, two more heads would grow in its place. If present day hydras were as large as the mythical ones, we too might consider them monstrous. They usually have five or six long wavy tentacles attached at one end to a hollow cylindrical body; but today's hydras are so small it takes a microscope to see even this detail. On the one hand they resemble plants, miniature trees with trunk and limbs, but lo and behold, the limbs stretch and twist and coil. It is an animal, and a carnivore at that. *Hydra's* diet consists exclusively of tiny animals that venture too close to the deadly tentacles. Thousands of special cells with threadlike projectiles, trigger-set to explode at passing prey, are in the tentacles and around the mouth. Some of these tiny threadlike projectiles wrap around the prey and entangle it. Others pierce the body of the prey, paralyzing it with poisonous secretions. This is truly a well-armed carnivore.

Just as on land some animals burrow into the soil, so too there are burrowers in the bottom sediments. Most of these organisms are either true worms or worm-like in appearance (Fig. 3.9). Tiny roundworms, called nematodes, wriggle their needlelike bodies through the mud or debris searching for protozoans and other worms as food. These animals, as well as the true segmented worms, must be adapted to the very low oxygen levels on the bottom. *Tubifex* (the sludgeworm) is a particularly well-adapted segmented worm. It can tolerate the extremely low oxygen levels in polluted waters where bacteria are abundant. *Tubifex* builds a tube above a burrow extending down into the bottom materials. While it feeds, mouth down in the bottom, its tail sticks out of the tube. It sways back and forth, exchanging carbon dioxide for any available oxygen in the water.

Another common bottom burrower is the bloodworm (Fig. 3.10). Its blood is often similarly colored to our own. The red pig-

Fig. 3.10
The bloodworm, *Chironomus*, is the larva of the midge fly, an annoying pest that resembles a small mosquito but does not bite.

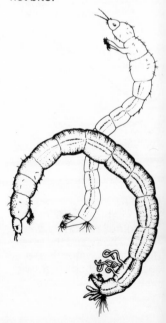

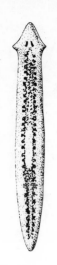

Fig. 3.11
The planarian, a representative flatworm.

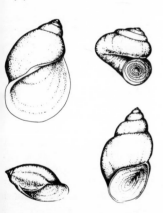

Fig. 3.12
Some of the many types of snails that you will find in ponds and small lakes.

Fig. 3.13
Crayfish occupy the niches of scavenger and predator in ponds.

ment aids the organism in breathing at very low oxygen levels, even on the bottom of deep lakes. Although it is worm-shaped, it is not a worm at all, but the larva of an insect. The insect is called a midge, and the adult form closely resembles a mosquito. Feeding on plankton and detritus, the larva, in turn, is often included in the diet of large and small fish.

The planarian (Fig. 3.11) is a favorite animal for laboratory studies conducted by biologists, teachers, and students alike. These intriguing creatures are common in lakes, ponds, pools, ditches, and streams. They avoid sunlight so are usually found under rocks and bottom debris. Here they feed on living and dead animal matter such as protozoa and nematodes. If it weren't that their dark brown coloration usually fits in perfectly with their environment, more people would know where planarians live in nature.

The remaining animal types are a mixed assortment of reasonably mobile foragers, scavengers, and predators.

Snails (Fig. 3.12) are normally found foraging for algae that often coat the bottom detritus and the leaves and stems of aquatic plants. Although it is hard to assess the importance of their role in the lake or pond, anyone owning an aquarium knows that they perform an excellent cleaning service.

Marine lobsters, shrimps, and crabs belong to a group of organisms called crustaceans. Just as insects dominate in numbers the land environment, crustaceans dominate the aquatic environment. Quite a variety of freshwater crustaceans exists in ponds and lakes. Four types are noted as active scavengers on the bottom.

Crayfish (Fig. 3.13) resemble miniature lobsters. They have a multitude of legs and other appendages for various functions. In a normal day's activity, a crayfish's walking legs might take it a few yards from its favorite den in search of food. Its large fighting claws might never be used, but then, if a larger crayfish comes along, one claw might be lost in battle. If this happens, the crayfish cannot start regenerating a new claw until its next molt.

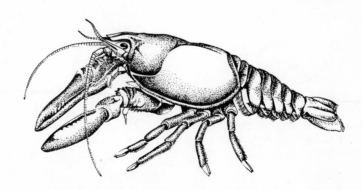

While scavenging, its mouth parts and appendages pick up and sample thousands of items on the bottom. Some are rejected, some consumed.

A second scavenging crustacean is the isopod (Fig. 3.14), commonly called the aquatic sow bug. The meaning of its name is quite simple. The prefix iso- means "equal" and the suffix -pod means "foot." All but the first pair of legs are, in fact, very similar. The terrestrial sow bug (pill bug or roly-poly) is the tiny animal often found beneath rocks or in other dark moist places. The aquatic sow bug prefers a similar habitat. It is fairly secretive, hiding and feeding beneath rocks, vegetation, and other bottom debris.

The third member of this group has the odd habit of swimming on its side. Its common name is scud or sideswimmer (Fig. 3.15). Its proper name is amphipod ("feet all around"). Scuds avoid bright light and feed most actively at night. However, when the vegetation is disturbed, they can easily be seen streaking off into the shelter and safety of the bottom debris.

Last and least in size is the ostracod or seed shrimp (Fig. 3.16). Barely visible to the naked eye, they look like suspended sediments in the water. They often occur in great numbers on and just above the bottom. There they feed on bacteria, molds, and algae. Unfortunately, they do not have the semi-transparent shells found in many tiny crustaceans. Their shells are pigmented, making it impossible to see the internal anatomy of a living specimen under the microscope.

Despite the fact that leeches (Fig. 3.17) are powerful swimmers, we must include them as bottom dwellers. They move about a great deal, but this chiefly occurs under the cover of darkness. At night, leeches glide over the bottom and through the vegetation either scavenging or looking for a blood meal. During the hours of daylight they usually hide beneath stones, vegetation, and other bottom material. Your field studies will be during daylight, so look for leeches in hiding.

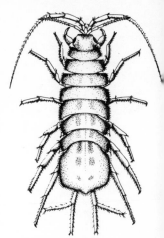

Fig. 3.14
Asellus is a genus of isopod that you will commonly find in ponds. This organism is about 1 cm long.

Fig. 3.15
The amphipod, like most crustaceans, has a large number of appendages with a wide range of form and function. Magnified 8X.

Fig. 3.16
Ostracods resemble tiny clams. Examine one under a microscope to find out why it is classed as a crustacean.

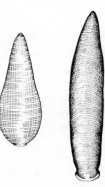

Fig. 3.17
Two of the many types of leeches that dwell in ponds.

3.3   IN THE BOTTOM VEGETATION

Certain creatures of our inland waters spend most of their time in the thick growth of aquatic vegetation in the littoral zones. These animals could be classed as bottom dwellers but, since they are usually associated with large aquatic plants, we shall place them in this microhabitat. Remember, though, that these organisms can sometimes be found on the bottom while, vice versa, snails, amphipods, isopods, and hydras are often found on the plants.

To see that some types of life do reside in more than one type of habitat, we need look no further than the mayflies. The mayfly is an insect. It spends most of its life in the water, leaving as an adult (Fig. 3.18), sometimes for only a few hours, to mate, lay its eggs, and die. All species of mayflies are herbivores, eating algae and larger aquatic plants. The adults eat nothing at all. Certain species of mayflies (Fig. 3.19) are found only in lakes, ponds, and quiet backwaters of streams. Others are found only in running waters. Among the still water types, some are found only in areas having submerged aquatic vegetation. Others are sprawlers on the bottom, and still others are burrowers in the bottom. Each type is

Fig. 3.18
An adult mayfly.

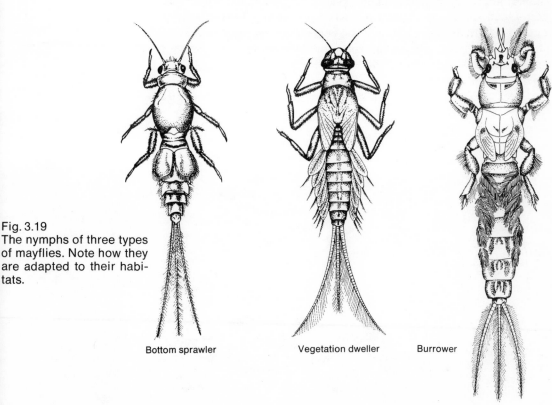

Fig. 3.19
The nymphs of three types of mayflies. Note how they are adapted to their habitats.

Bottom sprawler   Vegetation dweller   Burrower

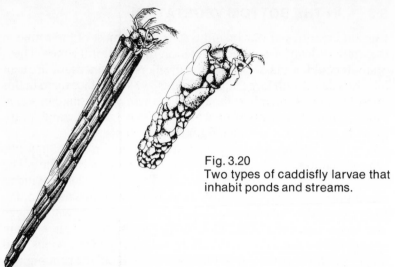

Fig. 3.20
Two types of caddisfly larvae that inhabit ponds and streams.

Fig. 3.21
The larva of the whirligig beetle.

specifically adapted to the particular microhabitat in which it is most often found. The environmental conditions within each habitat are different and demand specific capabilities in those species which survive in them. For instance, the mayfly species that climb and run and dart about in the vegetation have specially constructed gills and tail appendages for swimming. The bottom sprawlers show different modifications for living in the conditions on the bottom. The first pair of gills often covers the remaining gills and protects them from getting clogged with bottom silt. The bur-

Fig. 3.22
The dragonfly.

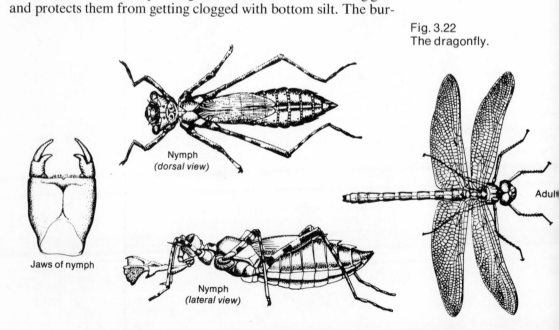

Jaws of nymph

Nymph
(dorsal view)

Nymph
(lateral view)

Adult

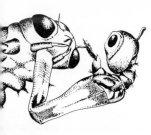

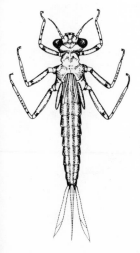

rowing species have jaws that plow through the mud, sand, and sediments. Special gills maintain a flow of oxygenated water over their bodies. You should examine the mayflies collected in your field studies for adaptations to the habitats in which you find them.

Caddisflies (Fig. 3.20) are insects having a four stage life cycle—egg, larva, pupa, and adult. All stages but the last live in the water. Those species living in the vegetation display an interesting tactic which may make it difficult for you to find them. They camouflage themselves by building a portable house of twigs or vegetation to match the surroundings. Pieces of vegetation are carefully cut and shaped before being bound together with silk. The end product is a stick that crawls!

Gyrinus, you will recall, is the whirligig beetle. Unlike the adult stage, the larva (Fig. 3.21) is independent of the surface. It lives in submerged vegetation. Like other aquatic beetle larvae, it is a fierce predator. It can crawl through the vegetation, swim in open areas, and kill with its large curving jaws. The sharp pointed jaws not only grasp and puncture the prey, but also drain the body fluids from it. They contain canals which lead to the mouth, and the fluids are sucked up as if through two straws.

During the late spring, summer, and fall you will surely see the aerial acrobatics of the adult dragonfly (Fig. 3.22). Swooping over the water in pursuit of a mosquito or fly, it carries out a definite service for man. Killing other insects is a lifelong occupation for the dragonfly. The aquatic nymph stage is as formidable a predator as the adult. With careful steps it travels about in the vegetation, stalking its prey. Even minnows are not safe from its gaze. Small worms, crustaceans, and mollusks fill out its diet. With the weaponry at its disposal, it's no wonder it has such varied meals. A special appendage is hidden beneath its head. When an organism comes into range, the appendage unfolds, shoots forward, and two tooth-like spikes sink into the sides of the prey (Fig. 3.23). Thus impaled, the meal is brought in for the kill.

The damselfly nymph (Fig. 3.24) feeds in a similar fashion. It is a close relative of the dragonfly. The two animals are quite different, though, in their modes of swimming and breathing. The damselfly nymph has three large terminal gills not found on the dragonfly nymph. The gills are, of course, important in breathing but they are also used as fins when the nymph swims in its awkward wriggling motion. In contrast, the dragonfly nymph has a large chamber in its abdomen instead of fin-like gills. Water drawn in and out of this chamber aerates tiny gills inside, aiding breathing. By contracting the abdominal muscles and forcing water out of the chamber, the dragonfly nymph swims using jet propulsion.

3.4 TINY SWIMMING AND DRIFTING PLANTS: PHYTOPLANKTON

If you were to compare plant life in water with that found on the land, you might do it as follows. The large rooted aquatic plants like cattail, pond lily, and bladderwort can be compared to the trees on land. These plants, for the most part, supply shelter and protection and thus provide a habitat; but they are not primarily used as food themselves. Tiny aquatic plants might be compared to the grasses on the land. As the grasses and weeds are important as food sources for the herbivores on land, so, too, are these planktonic plants in the lake or pond.

One of the simplest organisms structurally is the diatom (Fig. 3.25). Diatoms are minute, single-celled plants that sometimes occur in tremendous numbers early in the spring as part of the suspended or drifting life of pond or lake water. They possess two shells with intricate markings. The shells fit like the top and bottom to a box, with the cell protoplasm, the living material, inside. Some varieties have circular shells shaped like petri dishes. Others are boat- or needle-shaped. Under the microscope, close observation reveals the characteristic markings on the shell indicating that the organism is a diatom.

Ecologically, the diatom is important as the major component of the "grass" or phytoplankton, the basic food source for all consumers. When their numbers become very large, they may change the water's color. This usually happens one or more times during the year when temperature and sunlight are most suitable for growth and reproduction. Unicellular plant cells divide two to ten times in a month under proper conditions, each time doubling their numbers. At the higher rate of division, a single cell could form a population of 10^{15} individuals in a single season. Luckily there are a few herbivores around!

Desmids are another form of unicellular phytoplankton. On first examination one would think that desmids always occur in pairs (Fig. 3.25). The two identical halves have been called semicells, but both parts together constitute a single cell and a single plant. Why a plant should show such intricate symmetry is a perplexing question. Because of their small size, a fine-meshed plankton net is needed to collect them. This is also true with diatoms.

A desmid is just one variety of a diverse group of organisms called green algae. Many of the other varieties are also planktonic. Some are unicellular and motile (able to move) like *Euglena* and *Chlamydomonas* (Fig. 3.25). Others are colonial like *Volvox*, which consists of as many as 50,000 cells held together in a spherical jelly-like mass no larger than a pinhead.

Fig. 3.25
Some common phytoplankton of ponds and lakes.

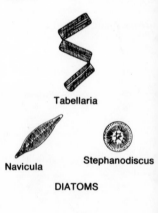

Tabellaria

Navicula

Stephanodiscus

DIATOMS

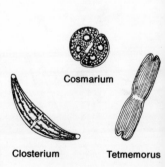

Cosmarium

Closterium

Tetmemorus

DESMIDS

In recent years another group of phytoplankton have become important, the blue-green algae (Fig. 3.25). Certain members of this group are often responsible for the so-called "algal blooms" in nutrient-rich and polluted waters. (Algal blooms are massive summer populations of algae.) They may give the water an appearance of pea soup and produce an unpleasant smell and taste. When the algae die, the oxygen depletion resulting from bacterial growth may kill off fish and further upset the ecosystem. Blue-green algae are often good indicators of the aging of a lake, and these and other phytoplankton should not be overlooked in your lake studies.

When a lake is first created, let's say from the scowering action of a glacier, its water may contain minute traces of certain chemical compounds. If compounds like nitrates and phosphates are present in small amounts, the lake will support only a sparse plant community. Such a lake is called *oligotrophic*, meaning unproductive. The other extreme is a lake with high concentrations of all the compounds needed for life processes. Such a lake is called *eutrophic*, meaning productive. With time, because rivers continually add compounds from the areas they drain, oligotrophic lakes age and become eutrophic. This is called *eutrophication*. Phytoplankton are an indicator of the aging or eutrophication that has occurred in a lake. Desmids dominate the phytoplankton in very oligotrophic, young lakes but are replaced by a predominance of diatoms under less oligotrophic conditions. As aging proceeds, flagellates and other green algae eventually dominate. Under final eutrophic, mature, conditions the blue-green algae are most abundant.

The next link in the many food chains that begin with the phytoplankton is usually the zooplankton—tiny swimming and drifting animals.

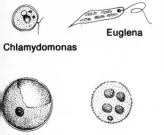

Chlamydomonas

Euglena

Chlorella

Volvox

OTHER GREEN ALGAE

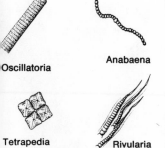

Oscillatoria

Anabaena

Tetrapedia

Rivularia

BLUE-GREEN ALGAE

3.5 TINY SWIMMING AND DRIFTING ANIMALS: ZOOPLANKTON

Three main groups make up the zooplankton—protozoa, rotifers, and crustaceans. Other zooplankton are rarely collected using ordinary methods so only a passing reference will be made to them.

Protozoa have already been classed as bottom dwellers. Some forms, however, are planktonic and will be collected in plankton samples. It is impossible to predict the particular species that you may find. Your best bet is to check a reference book like Needham and Needham (1962) to identify the various types when the time arises.

Rotifers are tiny multicellular animals which occur in nearly all freshwater habitats (Fig. 3.26). Of the 1,700 described species, 95% are restricted to fresh water (as opposed to salt water). Approximately 75% of the species can be found in the littoral zone of ponds and lakes. These are the areas most accessible to you. You can expect to find 40 to 500 rotifers in a liter of water from a normal plankton community. These are far fewer than the phytoplankton present, but the rotifers also play an important role. Like miniature lawnmowers, they keep the "grass" in check. Under the microscope they appear to have a pair of cutting wheels, similar to a lawnmower. Actually the wheels are synchronized beating rings of cilia that bring food particles to the rotifer's mouth.

Two important planktonic crustaceans are *Daphnia*, a common cladoceran or water flea, and *Cyclops*, a common copepod (Fig. 3.27). These crustaceans feed by filtering smaller plankton from the water. Although related to the crustacean scavengers on the bottom, the relationship is a distant one; they appear much unlike crayfish or sideswimmers. Like rotifers, these animals are usually numerous and ecologically important in food chains.

Two rather queer looking animals should also be described, if only to avoid a few anxious moments of disbelief should you find one of them. An animal that most everyone associates with the ocean is the jellyfish. But there is a "freshwater jellyfish" (Fig. 3.28). Although it is quite small and is really a relative of the *Hydra*, it has nearly everything that makes a jellyfish a jellyfish. Most people have never seen one, since they are somewhat rare. They are sometimes found in ponds, small lakes, and old flooded quarries. The other odd-looking creature is the phantom midge larva (Fig. 3.28). It is one of the largest zooplankton, sometimes over half an inch in length. It has the name "phantom" because of its ghostly appearance. Except for two silvery air sacs used to raise and lower it through the water, the beast is transparent.

Fig. 3.26
These are a few of the many species of rotifer that live in ponds.

3.6 AIR-BREATHING AQUATIC INSECTS

It is generally true that large consumers are much less numerous than small ones. The reason is that a given amount of habitat can support only a few large organisms which use a lot of food and energy. The same amount of habitat supplying the same amount of energy in the form of food can maintain many small consumers since they require less food individually.

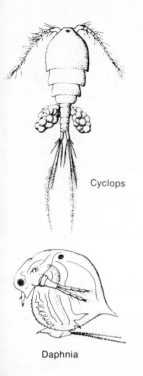

Cyclops

Daphnia

g. 3.27
vo common planktonic
ustaceans that live in
onds.

While searching in a lake or pond, you will find that the two factors of size and number compensate one another. Something large should be spotted easily even though there are few individuals. More numerous small organisms may be overlooked. Large aquatic insect life seldom goes undiscovered by an efficient worker.

You might be scared out of your wits, but, if you come across a giant water bug (Fig. 3.29), keep calm; you will have found a real treasure. It is less impressive to talk about "the one that got away" than to actually take it back to the classroom. You can then enjoy its antics for weeks or months.

The giant water bug may grow to three inches in length and an inch or so in width. It feeds on aquatic animals large enough to make a decent meal, for example, dragonfly or damselfly nymphs, tadpoles, or small minnows. As with all "true bugs," it has a jointed piercing and sucking beak with which it dispatches its prey. It also seems to have a toxic secretion which it can inject with its beak. Take care, then, if you collect this fellow by hand.

Only *Lethocerus* grows to the size mentioned; the more common *Belostoma* is about a third that size. The female of this species has the interesting habit of laying her eggs on the back of the male, who is obliged to protect them. When the eggs hatch and the not-so-giant water bug emerges, a fascinating change occurs. As its legs unfold from the tight egg case, the body of the bug grows rapidly, and in less than an hour is puffed up over five times the size of the egg case. One can imagine the tremendous empty feeling inside the young water bug. No wonder it immediately attempts to fill its very empty stomach.

Fig. 3.28
Two unusual inhabitants of ponds.

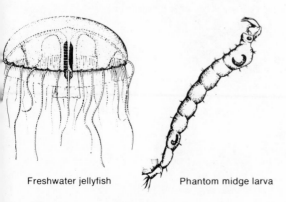

Freshwater jellyfish   Phantom midge larva

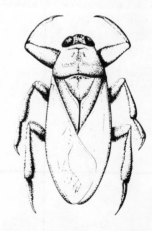

Fig. 3.29
Lethocerus, a giant water bug.

All of the insects discussed in this section are air breathers. The giant water bug must surface occasionally for air, as must *Ranatra* (Fig. 3.30), another true bug. Commonly called the water scorpion, *Ranatra* is a long, thin, gangly creature. It breathes through a straw formed from slender filaments extending from the tip of its abdomen. Hanging upside down, with the end of this tube out of water, *Ranatra* clings to the vegetation waiting for food. Its front legs respond much like those of the terrestrial preying mantis and lash out to seize passing food items. Both *Ranatra* and the giant water bug are found in ponds and weedy sections of lakes or streams. Here, *Ranatra's* stick-like shape and coloration give it excellent camouflage.

A close relative of the phantom midge is that nuisance insect, the mosquito (Fig. 3.31). The egg, larva, and pupa stages in the life cycle are aquatic. Both larva and pupa usually hang from the surface tension, obtaining oxygen through special respiratory tubes. When disturbed, often merely by a moving shadow, they swim in a violent wriggling manner down into the safety of the bottom weeds.

The mosquito larvae and pupae feed on minute algae and protozoa. They are, in turn, eaten by many carnivores including fish, damselfly nymphs, dragonfly nymphs, diving beetles, water boatmen, and backswimmers.

The predaceous diving beetle adult and larva (Fig. 3.32) are both extremely active killers. So fierce are the larvae that, if two equal-sized individuals are placed in a container, they are locked in deadly combat within a few minutes. They are aggressive and instinctively attack movement. Their large pointed jaws pierce the armor of most other aquatic animal life. The jaws are so large that a full-sized larva can stab through the "house" of twigs or other vegetation made by the caddisfly, kill the larva within, and suck out its body fluids.

Fig. 3.30
Ranatra, the water scorpion.

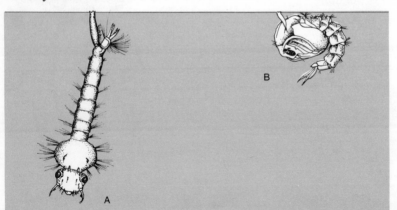

Fig. 3.31
Mosquito larva (A), an pupa (B).

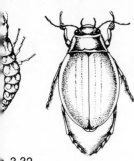

. 3.32
e larval and adult forms
Dytiscus, one of the
edaceous diving bee-
s.

The water boatman and the backswimmer (Fig. 3.33) are also active, air-breathing insects. Dangling from the surface, waiting for some unfortunate morsel to appear, the two animals look much alike. Each has long powerful swimming legs, a pointed beak, large wrap-around eyes, and roughly the same shape and size. The difference is the manner in which they swim, the backswimmer preferring to travel upside down.

You have now met a large number of animals. They have been grouped by their most common locations, to guide you in the field. Many of them, however, will be found in other habitats. For example, the seed shrimp was classed as a bottom dweller, but it is often captured with drifting and free-swimming plankton. The following chart gives you a more complete guide.

. 3.33
e common species of
e water boatman (on the
getation) and the back-
immer (hanging from
e surface) are about 10–
mm in length.

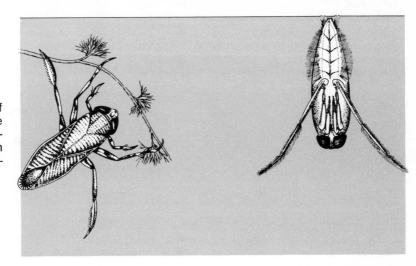

BLE 1

	On the surface	Hanging from the surface	Free swimming (□ occasional)	On or among vegetation	On the bottom	In the bottom sediments
ringtail / ater strider	■ ■					
antom midge / eshwater jellyfish			■ ■			
ussel / udgeworm						■ ■

(continued)

Table 1 (contd.)

	On the surface	Hanging from the surface	Free swimming (□ occasional)	On or among vegetation	On the bottom	In the bottom sediment
Whirligig beetle (adult)	■		□	■		
Fisher spider	■			■		
Water boatman		■	□	■		
Backswimmer		■	□	■		
Giant water bug		■	□	■		
Water scorpion		■	□	■		
Predaceous diving beetle		■	□	■		
Mosquito larva and pupa		■	□	■		
Crayfish			□	■	■	
Dragonfly nymph			□	■	■	
Damselfly nymph			□	■	■	
Whirligig larva			□	■	■	
Sideswimmer			□	■	■	
Isopod				■	■	
Bryozoa				■	■	
Sponge				■	■	
Bloodworm				■		■
Hydra		■		■	■	
Snail		■		■	■	
Planaria					■	■
Ostracod					■	■
Leech			■	■	■	
Copepod			■	■	■	
Cladocera		■	■	■	■	
Mayfly nymph			□	■	■	■
Caddisfly larva				■		
Nematode				■	■	■
Rotifer			■	■	■	■
Protozoa		■	■	■	■	■
Bacteria	■	■	■	■	■	■

3.7 LARGE ANIMAL LIFE

We have discussed the great numbers of small organisms present in most ponds and small lakes. It is not surprising, then, that some larger animals, taking advantage of this food supply, also live in lakes and ponds. A few of these larger organisms may occur in your study area.

Fish, if present, are always the most influential of the larger animals. They feed almost exclusively on other aquatic life. Thus they directly control the numbers of many other animal species. In a lake study though, it is difficult to sample the fish population. Without a very large net, these fast-moving animals are difficult to trap. With proper equipment and technique you might catch quite a variety of fish such as the following:

1) *Lepomis*, the "sunfish" genus, contains a number of species. The bluegill, pumpkinseed and green sunfish all interbreed, giving a variety of colorations.

2) *Notropis*, and other members of the family Cyprinidae, are commonly called "minnows." The common names of these fish, such as dace, chub, minnow, and shiner, have little or no taxonomic meaning.

3) *Micropterus*, the bass.

4) *Perca*, the perch.

5) *Esox*, the pike.

These common varieties of fish are shown in Figure 3.34.

The particular species found will be governed by the location of the lake or pond, its physical and chemical properties, available food, and predators (usually other fish species).

To properly identify a given specimen, you will need an identification guide designed for your area. A couple of possible books are listed at the end of Unit 3.

Most freshwater habitats have at least one or two types of amphibians present. Over the years, frogs have taken quite a beating from man, especially those thoughtless "frog hunters"—little boys that prowl the neighborhood ponds. Needless massacres occur, perhaps because, when in danger, frogs remain motionless until the last fateful second, counting on their camouflaged bodies to protect them. The newt is another amphibian, less well-known because it is rarer and more secretive than the frog.

Frogs show a marked change in diet from tadpole to adult. Tadpoles are mainly herbivorous, feeding upon algae, diatoms, desmids, and decaying aquatic plants. The intestine of the tadpole

Green sunfish

Pumpkinseed

Bluegill

Horned dace
(*creek chub*)

Blacknose dace

Common shiner
(*redfin*)

Yellow perch

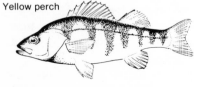

Northern pike

Largemouth bass

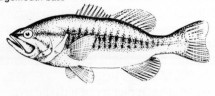

Fig. 3.34
Some of the fish that com
monly inhabit ponds an
small lakes.

is long and coiled as is typical of vegetarians. The adults are, in contrast, carnivorous.

A terrarium can be set up to keep a frog or two back in the lab. Keep in mind that frogs, newts, and other animals survive best in their proper environments, the ones that they select.

Turtles are familiar aquatic inhabitants, especially in mature ponds or marshy sections of lakes. They are true scavengers, feeding on almost anything. Depending on the species, their diet may consist of more than 50% vegetation. If you are fishing in snapping turtle waters, never leave a string of fish in the water unattended. These turtles speedily find dead or injured animals and devour them. Although they are rather ugly beasts, they carry out a needed service—keeping the water clean.

Muskrats can have a profound effect on an aquatic area, destroying the homes of many other organisms. Active vegetarians, they can strip areas of the pond below the water level and can hold back the natural process of succession, preventing the growth of emergent plants. Cattails, for example, are a preferred food. A muskrat den is recognizable as a pile of cattail stalks. Check the surface water around the den for uprooted and partially eaten plants. They will give an indication of the muskrat's food habits.

All sorts of waterfowl use large and small freshwater areas. Coots, grebes, and ducks are important top level consumers for a great many food chains.

Many of the larger animals are very secretive. It is hard to get a true idea of their importance in the ecology of a given study area.

3.8 LARGE AQUATIC PLANTS

Try to visualize the following situation. You are fishing from a boat anchored in a shallow section of a large lake. A friend is with you and you are hoping to land some good-sized largemouth bass. Having learned some ecology, you know that this fish lives in and around fairly shallow weed beds. In this habitat, a largemouth bass feeds on minnows, frogs, and aquatic insects and crustaceans. You are using some small crayfish for bait, since these are an important food for bass. So far you have impressed your dejected friend by catching the only fish. But your casting is not as effective as your ecological insights. On the next cast your hook and bait go right into a submerged bed of "weeds." Your friend looks over with a half smile and begins chuckling to himself. Finally, you have miscued . . . he thinks.

After retrieving the hook, bait, and collection of plant life, you do something rather odd. You examine the "weeds," turning them over in your hands, looking at the leaves and stems. Then, in an off-hand way, you toss them back into the water, commenting, "Hmmm, looks like sago pondweed to me. Yes, it must be good old *Potamogeton pectinatus*; couldn't be anything else. You do know, of course, that it's a very important food for most species of ducks? This part of the lake should be a great place to hunt in the fall. The ducks probably drop in here by the dozens while on their migration. We must come back someday."

By this time your friend may have fallen right out of the boat in amazement. But aquatic plants *are* as different from one another as pine trees from oaks. They have specific names and don't have to be referred to merely as "weeds." Interestingly, probably over 75% of the aquatic plants snagged by fishermen belong to just five or six common types, each with an impressive name like *Myriophyllum* or *Ceratophyllum*.

Some knowledge of types of aquatic plants will be important to you in your studies of a lake, pond, or stream habitat. To understand the ecology of the pond or lake you must realize two points:

1) Large aquatic plants modify the habitat by their presence. They furnish a variety of microhabitats important to aquatic animal life.

2) The various species of large aquatic plants are restricted to given habitats. The competition of other aquatic plants and their own tolerance to conditions like bottom type, water movement, and amount of sunlight (depth and turbidity) are factors here.

Visualize yourself still in the boat with or without your friend. Now, what can you expect to find and where? If the water gets deep quickly at the edge of the visible "weed" beds, the anchor may have landed on a dark bottom, free of aquatic plants. Without light, green plant life cannot exist. If, however, the water was not very deep, the anchor might be resting in a luxuriant green mat. Pull the anchor up and you will get specimens of *Myriophyllum* and *Ceratophyllum* (Fig. 3.35). The latter plant, commonly called coontail, is distinguished from the former by a serrated edge along its narrow leaves. Both plants are of little use to wildlife as food. They can, however, tolerate changing water levels and moderate turbidity. Thus, in some areas they may crowd out more desirable plants.

Fig. 3.35
Common plants that live on the bottom of ponds.

Chara
(muskgrass or stonewort)

Utricularia
(bladderwort)

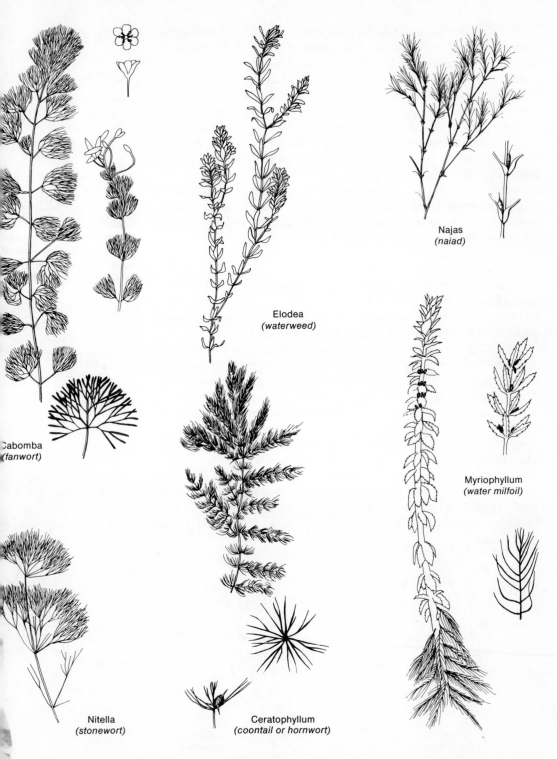

Najas
(naiad)

Elodea
(waterweed)

Myriophyllum
(water milfoil)

Cabomba
(fanwort)

Nitella
(stonewort)

Ceratophyllum
(coontail or hornwort)

Large Aquatic Plants   53

Other vegetation that could come up with the anchor includes *Chara*, *Nitella*, *Najas*, *Cabomba*, *Elodea*, and *Utricularia* (Fig. 3.35). *Chara* and *Nitella* are algae and, as such, are more primitive life forms than a plant like *Najas*. *Elodea* (common name waterweed) is best known as an aquarium plant. Both *Elodea* and coontail live quite well in aquaria, even when unattached to the bottom. Presumably they get minerals directly from the water.

Utricularia, commonly called bladderwort, is a carnivorous plant. As its common name suggests, it has bladder-like traps (Fig. 3.36) among its leaves. The bladder is normally sealed shut by a valve or watertight door at one end. Small glands in the trap expel water from inside, and the walls of the bladder become concave because of the decreased pressure. On the outside, around the trap door, are a number of hair-like structures. Some act as a funnel to channel animals towards the door. Others act as a trigger device. When a small crustacean or other organism in the right position touches one of the trigger hairs, the valve opens. The walls spring back to their original shape and suck water into the bladder. Of course the prey is normally sucked in also. Quickly the valve goes back into place and the animal's fate is sealed.

With the anchor up and with a favorable wind, the boat may get carried towards the shore. Just looking over the side, you should recognize some of the plants brought up with the anchor. *Myriophyllum*, commonly called water milfoil, may grow to within inches of the surface. Coontail and bladderwort will do the same. In late summer, they send flowers up above the water level where wind and flying insects aid in pollination. You may also spot some different plants. The ribbon-like leaves of wild celery, *Vallisneria* (also called eel grass), reach up close to the surface also (Fig. 3.37). The female flowers of this plant are separate from the male ones. They float on the surface connected by a long, thin filament to the base of the plant. The male flowers break off from the base of

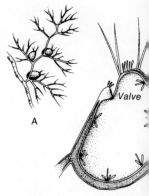

A

Valve

Fig. 3.36
The bladderwort, showir
(A) a single leaf bearir
traps, and (B) a longitud
nal section of a trap.

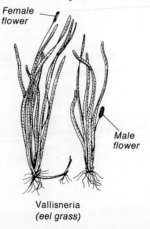

Female
flower

Male
flower

Vallisneria
(eel grass)

Potamogeton
(pondweed)

Fig. 3.37
These plants common
reach up to the surface
the water.

the plant and rise to the surface. They float freely while liberating their pollen. When the female flowers are fertilized, the filament coils up like a spring and pulls the female flower below the surface, out of sight, to produce seeds.

Inevitably, your boat will drift through pondweeds (Fig. 3.37), the *Potamogeton* family. These plants are extremely important to waterfowl. The narrow-leafed sago pondweed, *Potamogeton pectinatus*, is the most valuable aquatic food plant to wildlife.

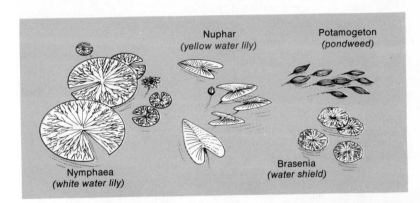

Fig. 3.38
Floating-leafed plants.

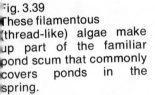

Spirogyra

Zygnema

It is eaten by coots, swans, geese, most species of duck, shorebirds, muskrats, and even moose. Every part of the plant—seeds, stems, leaves, and rootstocks (tubers)—is consumed by one animal or another. The broad-leafed pondweeds like *Potamogeton natans* usually have some leaves floating on the surface. This foliage is seldom eaten, but the seeds are a delicacy for ducks and geese in the fall. The floating leaves are often different in shape from the subsurface ones, and have a waxy upper surface that sheds water.

The surface leaves of the broad-leafed pondweeds are usually found with other floating-leafed plants (Fig. 3.38). The white water lilies, *Nymphaea*, and the yellow water lilies, *Nuphar*, should be present. With the rarer plant *Brasenia*, water shield, they may form great surface rafts of greenery. These completely shade out sections of the pond, keeping it dark and cool. The surface plants, incidentally, constitute another microhabitat. Many beetles and other animals live on or under the floating leaves.

In still shallower water, you may find bright green masses of threadlike material. Although most commonly found in stagnant ponds or pools, various types of algae (Fig. 3.39) will be seen. Since the strands are so narrow, you will need a microscope to distinguish between species.

On any stagnant area of a pond you are likely to find varieties of a fascinating group of plants called duckweed. One of the larger varieties, *Spirodela* (Fig. 3.40), is only one-third of an inch long. Its oval leaves float on the surface and its tuft of roots hangs below. Under the proper conditions, reproduction of this plant is very rapid. Indeed, in late spring and early summer it often covers ponds so densely that red-winged blackbirds can actually walk across the surface. Another variety of duckweed, *Wolffia* (Fig. 3.40), is of interest to botanists because, with a length of only 1/25 of an inch, it is probably the smallest of the true flowering plants. This plant is often missed on field trips because it exists as inconspicuous green dots among larger varieties of duckweed.

Spirodela   Wolff

Fig. 3.40
Two genera of duckweed.

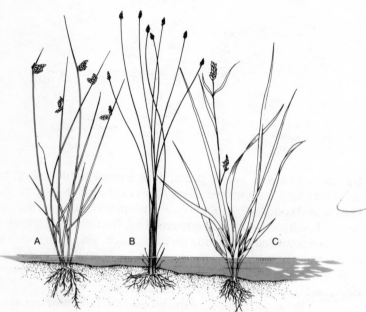

Fig. 3.41
The sedges like *Scirpu* (A), *Eleocharis* (B), an *Carex* (C), are all charac terized by triangular stem that are solid (not hollow like grasses).

Unless the wind is exceptionally strong, it won't take long before your boat comes to rest among the plants that stick out of the water nearest the shore. Tall slender plants include *Scirpus*, bulrush, and *Eleocharis*, the spike rush (Fig. 3.41). *Carex* is a sedge, which can grow in the water or inland among the shore types. Besides these narrow grass-like plants, you may find broadleafed species, some with beautiful flowers. *Pontedaria* (Fig. 3.42), commonly called pickerel weed, produces a large blue flower that sticks up above the water's surface among the lush foliage. *Calla*, water arum, displays an unusual white flower (Fig. 3.43). The arrowhead, *Sagittaria*, and *Alisma*, water plantain, are also water plants with white flowers.

Closer to the shore you should find many characteristic marsh plants. One wonders at this point whether plants like cattail,

Fig. 3.42
Pickerel weed.

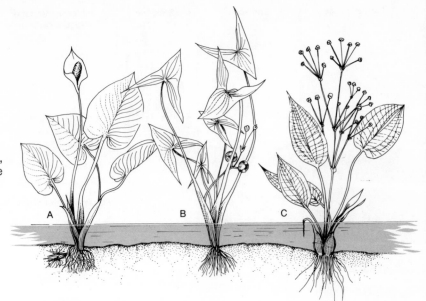

Fig. 3.43
Calla (A), *Sagittaria* (B),
and *Alisma* (C), all have
white flowers.

Fig. 3.44
The cattail, *Typha*.

Typha (Fig. 3.44), are really true aquatics or not. Little more than an inch or two of the stems of these plants may be in the water. Depending on the local conditions, *Typha* could be found rooted in six inches of water, floating in a mat on top of the water, or growing in the wet soils at the edge of the pond or lake, not submerged at all.

It would be a mistake to view and collect all these plants without recognizing the dynamic ecological phenomenon that binds them together. This is the phenomenon of succession.

Ecological succession refers to the successive habitation of a given site by different communities of organisms. Each successive community is better suited to the changing site conditions that the previous community may in fact have helped to create. That is quite a mouthful, so let us see by example what succession is.

If you have ever landed a boat at an undisturbed stretch of shoreline, you have probably noticed a zonal arrangement of the aquatic plants. You pass over a number of plant communities, each containing a different group of plant species (Fig. 3.45). Passing from the central, deep and dark, barren zone of a lake, you first encounter the zone of submergent aquatics. Plants like water milfoil and coontail that tolerate low levels of incident sunlight can exist in these deep areas. The submergent plants can be defined as those plants rooted to the bottom with leaves completely below the surface. Going into shallower water, you come to a different plant community, or zone, made up of surface aquatics. These are plants rooted in the bottom with their leaves floating on the surface. (You should recall these types, but if not, check Figure 3.38.) The last

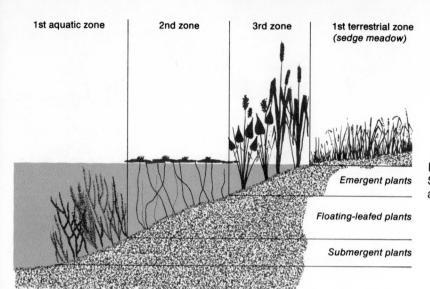

1st aquatic zone   2nd zone   3rd zone   1st terrestrial zone
(sedge meadow)

Emergent plants

Floating-leafed plants

Submergent plants

Fig. 3.45
Succession in space o
aquatic plants.

zone you pass through is that of emergent aquatics. The roots and part of the stems of these plants are located beneath the surface while their tops protrude from the water.

These various zones represent, in space, the succession of plant types that will occur in time at any spot in the lake or pond (Fig. 3.46). Why should this be true? As a result of deposits of sediments from the erosion of surrounding land, any area of the pond or lake will gradually become more shallow. Unfortunately we cannot plunk ourselves down near the edge of a pond, lie back, and

Fig. 3.46
Succession in time of aquatic plants for a given section of a lake or pond.

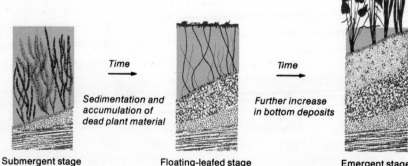

Time

Sedimentation and accumulation of dead plant material

Time

Further increase in bottom deposits

Submergent stage   Floating-leafed stage   Emergent stage

observe over a period of a couple of thousand years. If it were possible, we would see the bottom slowly rising towards the surface as sediments and dead plant material gave greater and greater thickness to the bottom deposits. At first, when the bottom was well below the surface, the deep water submergent plants would have grown in the area being watched. Hundreds of years later, with the bottom of the pond closer to the surface, the next stage in succession, the surface aquatics, would become established. In time, the water in the given section would be so shallow that emergent plants could take over and dominate as the next successional stage. Fortunately for us, all depths and their corresponding plant communities are usually visible at a single short visit to the lake or pond. We don't have to wait a thousand years for the next successional stage; we can see at a glance the plants that live at various depths. We thus can foresee that as deep areas become more shallow their present plant communities will be replaced by those of the next shallower zone.

Given enough time any pond or lake should fill in completely. The process of succession does not stop at this point, however. A whole series of terrestrial plant communities will in turn successively dominate the site. As before, the successional stages that will take place in time at a given site can be seen in space (Fig. 3.47). The next time you find yourself at the edge of a wilderness

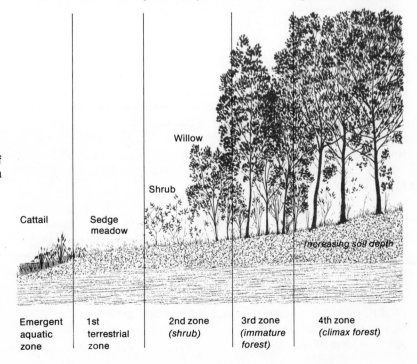

Fig. 3.47
Succession in space of terrestrial plants around a pond or lake.

lake, unspoiled by human beings, stand at the edge. With your toes submerged and your heels high and dry, note the aquatic vegetation zones.

If you were to make an about-face, you would find yourself looking at the shore plants. Their environment may be flooded for part of the year and very moist for the remainder. This is the point at which we paused originally to discuss succession. But succession doesn't stop here, so let's follow this process through to the end.

In the immediate foreground, maybe even touching your nose, you should see plants of the first terrestrial stage following aquatic succession. These plants are of the sedge meadow zone. They include types like cattail, *Typha*, the sedge *Carex*, and other slender plants like reed grass, *Calamagrostis*. This stage will, in time, be replaced by moisture-loving shrubs like dogwood, *Cornus*, and shrub willow, *Salix*. From the shoreline you should spot the bushes of this zone further inland.

The competition for sunlight inevitably results in trees of greater height succeeding the shrub types, once enough soil has been formed. Plants like alders (*Alnus*), birches (*Betula*), poplars (*Populus*), and spruces (*Picea*) reach up to capture the sun's energy needed for life. Smaller shrubs must give way. They are shaded to death. This third terrestrial stage, represented by a more inland zone, might be called the immature forest stage. It must not be forgotten that, throughout these stages, at any given site, the soil has grown in depth. Each successive community of plants has altered and added to the soil build-up. Eventually shade tolerant trees like maple (*Acer*) or beech (*Fagus*) get established. Their seedlings are capable of growing in the shade of the mature members of their own species. The particular species making up this mature, or climax, forest stage depends on the climate of the particular area. The important thing is that these tree types can perpetuate their own kind and remain the dominant plants at any given site, barring something catastrophic like a forest fire or a chain saw. The endpoint in successional development has been reached, the *climax* forest.

3.9   STUDYING INDIVIDUAL ORGANISMS IN DETAIL

Every organism has evolved in its own ways to meet the problems of obtaining energy for growth, avoiding death, and ultimately reproducing its own kind to perpetuate the species. Pond animals are no different in this respect from land animals. Basically there are two questions that we should always keep in mind when observing any organism. First, what physical or anatomical features does it

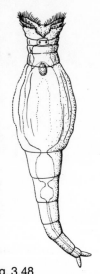

Fig. 3.48
What is it? How does it move? How does it obtain food?

possess? Specifically, what structures have evolved for breathing, locomotion, feeding, reproduction, and protection? Second, what types of behavior does it display? In other words, how does it respond to its environment (Fig. 3.48)?

Behavior can, of course, be seen only in the living organism and then only under close observation in reasonably natural conditions. We must look closely at individual animals to see the real secrets of their existence.

Looking at one aquatic animal as an example will indicate the amount of information that you can obtain through close observation.

A filter feeder. Interesting adaptations to pond existence occur in those animals which we may group as filter feeders. They have developed a special method of feeding upon the tiniest bacteria and other organisms suspended in the pond water.

When biologists search for microscopic organisms living in water, they always take along a plankton net. The mesh of such a net is so fine that organisms just barely visible to the naked eye are unable to pass through the pores. Some of the minute creatures caught will be filter feeders. They, in turn, have their own filtering system for catching even smaller life. As it turns out, the plankton net is very crude when compared to the filter feeder's apparatus.

Can you predict what sort of mechanism could have evolved to carry out this feeding procedure? In some cases, only organisms one-thousandth of a millimeter or less (1/25,000 of an inch) will pass through these natural filters. The filters hold back protozoans, algae, and even most bacteria.

The cladoceran, as its common name, water flea, implies, is a small flea-shaped animal found in most lakes and ponds. One of the most common genera is *Daphnia* (Fig. 3.27). The only freshwater areas where you are *not* likely to find water fleas are rapid flowing streams and badly polluted waters. Can you answer why? Although the general shape is similar to a flea, the organisms are unrelated. Their actual anatomy and ways of life bear this out.

Under the microscope, the tiny speck becomes a complex and intriguing little machine (Fig. 3.49). Many of its moving parts respond much as they would if they were still in the pond or lake. The relatively large antennae are used in swimming. With proper lighting the powerful muscles within the antennae may be visible. The body is encased in a shell which completely surrounds most of the other parts. It is transparent and allows one to see the internal structure. Five pairs of legs beat vigorously within the shell. Occasionally the whole posterior foot-like abdomen may strike out with its large terminal claw. The abdomen and claw can be used to propel the water flea through tangles of algae or bottom weed where it

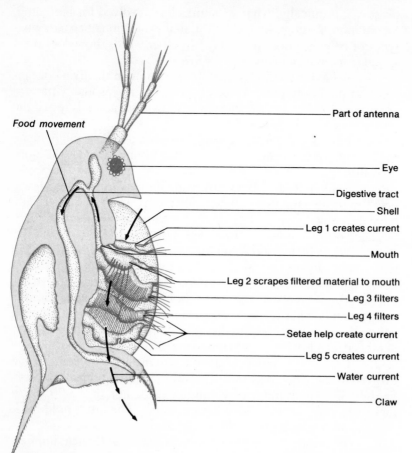

Food movement

Part of antenna

Eye

Digestive tract

Shell

Leg 1 creates current

Mouth

Leg 2 scrapes filtered material to mouth

Leg 3 filters

Leg 4 filters

Setae help create current

Leg 5 creates current

Water current

Claw

Fig. 3.49
Internal anatomy of *Daphnia*.

cannot swim. The five pairs of legs are also important because they are modified for obtaining food.

Looking at the water flea under the microscope, one often sees swirling currents of water when the legs begin to move. The feathery hairs at the tips of the legs create currents of water outside and inside the shell. These hairs, or filaments, are called *setae*. The all-important current is one that carries food into the shell. There it is trapped on the filtering setae of the third and fourth legs. Located on the second legs are bristles which occasionally scrape or comb material from the filters and draw it up to the mouth opening. See Figure 3.49.

As the water flea swims about the lake or pond, or clings to a plant or to bottom debris, the legs beat efficiently in harmony. They supply a continuous flow of food into the intestine, which is invariably full.

Filter feeders play an extremely important role in maintaining a balance in the pond. They help keep population levels of

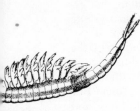

Fig. 3.50
The fairy shrimp, a resident of temporary ponds.

bacteria, algae, and protozoans in check. At the same time, they represent an essential link in food chains of insects, fish, and larger life forms.

Copepods are one-eyed little beasts similar in size to water fleas. Their antennae create a current and specially adapted mouthparts filter smaller plankton. *Cyclops* (Fig. 3.27) is one of the more common copepods that you will find. The fairy shrimp (Fig. 3.50), another many-legged animal, strains food out of the water with setae and concentrates it in a groove along the outside of the body. After a sticky binding material is added, the food is ready to eat. Completely defenseless, fairy shrimps are found only in small ponds and pools where no large predators exist. Adults live only for a short time in the spring, and have been known to hatch, grow to maturity, produce eggs, and die within a 15-day period.

Freshwater clams and mussels as well as some fish species are filter feeders. It is left to you to discover how they manage the task.

For Thought and Research

1 (a) What are the functions of the phytoplankton in a pond?
(b) What effect would a cloudy day have on these functions?
(c) Would very windy conditions affect these functions in any way?

2 Construct 5 food chains involving organisms discussed in this unit. Begin each food chain with a producer. If possible, draw a food web involving some or all of these food chains.

3 Trace the sequence of events that would likely occur in a pond if an algal bloom occurred. Would the pond community be permanently altered as a result of this bloom? Explain.

4 (a) What would happen to a pond community if all of the microorganisms were to suddenly die? Why?
(b) Imagine that conditions in a pond favor the production of unusually high numbers of planktonic herbivores like rotifers and copepods. What other changes will accompany this change? Account for your answer.

5 Visit a pond and collect enough duckweed to cover about 2 square feet of water surface. Place half of the duckweed in an aquarium full of water. Mount a 100 watt light bulb 1–2 feet above the aquarium. Do not aerate this aquarium (to duplicate natural conditions). Place the other half of the duckweed sample in an identical setup in which the water is aerated heavily. Observe both aquaria once or twice a week for several weeks. Describe and account for any observed differences. What zooplankton seem to use duckweed as a habitat? How do they respond to heavy aeration? Why?

6 Isolate and culture one planktonic pond organism (either phytoplankton or zooplankton). The techniques for doing this are outlined in *A Sourcebook for the Biological Sciences* by E. Morholt et al., Harcourt Brace Jovanovich, 1968. Keep a list of all of the things you learned about the organism during the course of this study.

7 Select one animal in the mini-ecosystem that you set up (Section 5.20). You may choose a protozoan, rotifer, cladoceran, copepod, isopod, amphipod, snail, sludge-worm, or any other animal that has survived for over a month in the mini-ecosystem. Do a detailed study of this animal. What food chains is it in? What niche does it occupy in each of these food chains? How is it adapted structurally to enable it to fill the niches that it occupies? What changes would occur if all of this type of animal died in the mini-ecosystem? Could a stable mini-ecosystem or natural pond exist without this type of animal? What chemical and physical conditions does the animal prefer? For example, what does it require in the way of cover, aeration, bottom conditions, and light intensity?

8 Select one of the many insect larvae and nymphs that are discussed in this unit. Research its life cycle using encyclopaedias and biology texts. Determine the niches occupied by the adult form of the insect. Would man be affected in any way if this species of insect were completely wiped out? If you can duplicate most of the natural conditions in which the insect larvae and nymphs are found, you can raise adults in your classroom. Try it. Don't forget to provide suitable food. For example, the nymphs of damselflies and dragonflies should be supplied with live prey like worms and insects.

9 (a) How can you recognize a pond or lake with a high productivity?

(b) A small lake had very few fish in it. The owner, therefore, stocked it with young fish. The growth of these fish was stunted. Why?

10 Collect, if possible, a live specimen of each of the air-breathing aquatic insects discussed in Section 3.6. Compare their habits, such as modes of breathing and swimming. What unique structural adaptations does each insect possess? What niche is occupied by each insect? How is each insect equipped to fill that niche? Keep the containers covered. Otherwise, many of these insects will fly away.

11 Perform some of the following laboratory studies now and the remainder after you have read Unit 4.

5.23 *Laboratory Investigation of Breathing Mechanism*
5.24 *Plankton Filtration in the Field or Laboratory*
5.25 *Laboratory Study on Animal Behavior*
5.26 *Classroom Experiment: Liebig's Law of the Minimum*
5.28 *Classroom Experiment: Habitat Selection*

Recommended Readings

The following three books are easy to read and contain very interesting material regarding the ecology in and around ponds and lakes. Full-color illustrations support the written material.

1 *The Life of the Pond* by W. H. Amos, McGraw-Hill, 1967.
2 *The Life of the Marsh* by W. A. Niering, McGraw-Hill, 1966.
3 *The Great Lakes* by R. T. Allen, Natural Science of Canada Ltd., 1970.

Consult the following books for the identification of organisms that are not discussed in this book.

4 *A New Field Book of Freshwater Life* by Elsie B. Klots, G. P. Putnam & Sons, 1966.
5 *A Guide to the Study of Fresh Water Biology* by J. G. Needham and P. R. Needham, Holden-Day Co., 1962.
6 *A Field Guide to the Reptiles and Amphibians* by R. Conant, Houghton Mifflin, 1958.

7 *Pond Life* by G. K. Reid, Golden Press, 1967.
8 *Fishes* by H. S. Zim & H. H. Shoemaker, Golden Press, 1955.
9 *Algae in Water Supplies*, U. S. Public Health Service Publication No. 657.
10 *How to Know the Aquatic Plants* by G. W. Prescott, Wm. C. Brown Co., 1969.
11 *How to Know the Freshwater Algae* by G. W. Prescott, Wm. C. Brown Co., 1970.
12 *How to Know the Protozoa* by T. L. Jahn & F. F. Jahn, Wm. C. Brown Co., 1949.
13 *Underwater and Floating-leaved Plants of the United States and Canada*, U. S.
Bureau of Sport Fisheries and Wildlife, Resource Publication 44, 1967.

Streams and Rivers

4

Our primary concern throughout is the relationship between the *biotic* (living) and the *abiotic* (non-living or physical) factors which make up the aquatic environment. This *dynamic* or ever-changing interaction is generally more noticeable in the constantly moving water of streams and rivers than it is in the still water of ponds.

As you stand on the bank of a clear, rapidly moving trout stream, or as you watch children at play on the banks of a warm, sluggish, center-of-town river, think about the millions of gallons of water moving past your feet. Where is the source of the flow? Where does the flow end? If you follow the muddy town river in the direction of its source, you may find that it is fed by the trout stream that you were fishing in yesterday. Why do rivers and streams undergo such drastic changes? What causes these changes to take place? There are marked differences among the types of plants and animals to be found along the course of the stream. You know, yourself, that you would not fish in a slow, muddy river and expect to catch a trout. Nor would you search through the clean gravel on the bottom of a fast stream for leeches, tadpoles, or mosquito larvae. You can guess that much of the life that exists in ponds, small lakes, marshes, and swamps just could not exist in rapidly moving

streams. Why should this be so? What are the unique characteristics of streams? A study of these characteristics is a necessary pre-
requisite to understanding how the biotic and the abiotic factors in-
teract. Let's take a look at them.

4.1 CHARACTERISTICS OF STREAMS

The most important characteristic of streams is the velocity of flow.
It controls many other factors of lesser importance. Factors such as
oxygen content, water temperature, composition of stream bed,
amount of food available, and pollution level are all influenced by
the speed at which the water is flowing. All of these factors, in turn,
directly affect the types and the numbers of plant and animal species which exist, that is, the productivity.

(a) **Rate of flow** itself requires adaptation by plants and
animals. Since the water is in constant, often quite violent, motion,
you would expect any life to be swept along by the current. Organisms which survive such conditions must have some way of secur-
ing themselves to the bottom material. The larval and pupal forms
of insects such as blackflies, riffle beetles, caddisflies, dragonflies,
mayflies, and stoneflies show many such adaptations (Fig. 4.1).
The blackfly and riffle beetle larvae have sucker-like structures on
their ventral surfaces. The cases in which some species of caddisfly
larvae live can be secured to the bottom material. This gives protection from the current. A flattened, streamlined shape allows the
nymphs of stoneflies and mayflies to adhere to the undersides of
rocks and avoid being swept away. In sandy or silt-bottomed
streams, many organisms such as mayfly and dragonfly nymphs are

Fig. 4.1
Adaptations of some
stream insects.

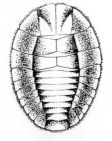

Water penny

Blackfly larva

Case

Point of
attachment

Caddisfly larva

Stonefly nymph

Mayfly nymph

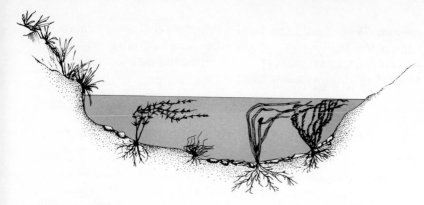

Fig. 4.2
Adaptations of some types
of stream vegetation.

partially submerged in burrows which they construct for their own protection. Similarly, plants are equipped with strong, sturdy roots and thin pliant stems and leaves which will not resist the current and break (Fig. 4.2). Many microscopic plants such as algae and diatoms remain fixed to a surface and offer very little resistance to the water flowing past them. Can you list other ways in which stream organisms have adapted to life in rapidly flowing water? Don't forget animals like fish, other vertebrates, crayfish, and clams.

(b) The **amount of oxygen** available to aquatic organisms is controlled to a large extent by the amount of agitation or aeration of the surrounding water. The turbulence of running stream water ensures an abundance of oxygen at all times. We expect, then, to find oxygen-loving organisms in this type of environment. Here respiratory processes can occur not only at the surface, but also at any depth below the surface.

(c) You will find, surprisingly, that although most narrow streams are shallow enough to allow penetration of sunlight, the **water temperature** is quite low. One factor which contributes to this is the heavy growth of vegetation shading the stream, preventing the sun from warming the water. The most significant cooling factor, however, is the rapid rate of evaporation. You perhaps know that stirring a liquid increases the surface area from which evaporation can take place. How does this mechanism operate in a stream?

You will remember that, in a pond, there is often a marked *vertical* temperature gradient (Fig. 2.4). From what you already know, what conclusions can you draw about a temperature gradient in streams? Since there is a constant turnover of water, the temperature will clearly be close to uniform throughout. The only change, then, will be *longitudinal* in nature. The velocity of flow is determined by the slope and shape of the land. Because the amount of aeration, evaporation, and thus cooling is influenced by this velocity, there will be a warming trend as the land gradient de-

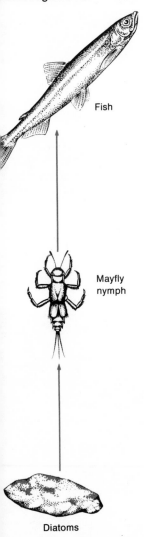

g. 4.3
quatic food chain show-
g relationships among
ream organisms.

Fish

Mayfly
nymph

Diatoms

creases. There will also be a longitudinal distribution of plant and animal species since only certain species are cold-tolerant. Be sure to watch for this on your field trips. We shall make further reference to temperature later when we discuss stream types.

(d) There is a direct relationship between **composition of the stream bed** and rate of flow. Each can affect and be affected by the other. A rapidly moving stream will carry along much of the debris, leaving only the larger and heavier particles behind. In slower waters, the bottom debris contains smaller and lighter particles.

In an area where a stream is in its beginning stages, the composition of the land greatly affects the rate at which the water flows. Large rocks and boulders resist the flow and thus slow the stream down. Smooth ground such as bedrock and very small particles offer little resistance. The composition of the stream bed will be discussed in greater detail later.

(e) The **amount of food** available in a stream is determined by all of the above characteristics. When producers are plentiful, as in areas with ample sunlight, consumers also flourish. Diatoms coating a rock feed primary consumers such as mayflies. They, in turn, feed higher order consumers like stoneflies and fish (Fig. 4.3). Overhanging vegetation also is a food source, supplying a variety of unfortunate terrestrial insects to the menu.

Stream organisms in fast waters have many specialized methods of obtaining their food. Since the material is continually moving past them, they must be able to grasp it quickly, or filter it out of the water while remaining stationary.

Food chains, both simple and complex, are easy to observe if you take the time. See how many different food chains you can list which include the organisms you see in a stream.

(f) The **pollution level** in streams drastically affects the productivity. Since dissolved oxygen is vital to all types of stream life, one of the most serious types of pollution is that which lowers the dissolved oxygen level. For example, the decay of organic matter such as sewage, wood chips, or dead algae consumes oxygen. If such pollutants are present in high concentrations, the oxygen supply can be seriously depleted. The result is a reduction in the productivity of the stream. Pollution is an involved topic. It is considered in great detail in another book in this series.

(g) **Other factors.** Many other physical and chemical factors must also be considered when you are investigating the ecology of streams (that is, the relationship between the biotic and the abiotic components). They include alkalinity, carbon dioxide content, hardness, pH, total suspended solids, and total dissolved solids. Reread Section 2.4 before performing your stream study.

4.2 VARIABILITY IN STREAMS

Streams originate from many different sources, flow over various types and slopes of land, and end in bodies of water as different as oceans and land-locked ponds. Springs, ponds, lakes, and seepage areas are starting points for many streams. Mountain streams consist mainly of *runoff* from precipitation and thus are generally active only in spring. This is vividly illustrated by the raging torrent which may flood an area in spring and then dwindle to a slow trickle by late summer. In any stream study, you must consider the amount of precipitation from season to season in the study area.

Each type of stream has a distinct set of *indicator (index) species*, plants and animals that thrive only under the physical and chemical conditions in that type of stream. Obviously, however, no two streams are exactly alike, and you may find very different organisms in two apparently similar situations. Any stream which you select, regardless of size or other characteristics, will provide an excellent opportunity for many field studies.

You will discover that a stream undergoes many changes along its course. These changes depend on the slope of the land, vegetation, amount of runoff, and many other factors. The *headwaters* (or source) of a stream may be fast-moving and turbulent, while at other points it may slow down and resemble a small pond. There may often be a series of small rapids or *riffles* interrupted at intervals by *pools* which hold the water for a short period of time. An interesting study is to follow a stream for as great a distance as possible and record all the distinct changes which occur along the way.

One of the most significant changes is in the composition of the bed. In a very fast-moving stream, the current carries away particles as large as ¼-inch in diameter, leaving nothing but stones on the stream bed. As the water slows down, finer particles settle out until, in extremely slow areas, the bottom is silty or muddy. At this point the stream possesses many pond characteristics. There are

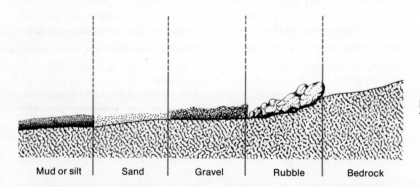

Mud or silt | Sand | Gravel | Rubble | Bedrock

Fig. 4.4
Types of stream beds.

five distinct categories of stream beds. These are *bedrock*, *rubble* (stones larger than gravel), *gravel* (coarse and fine), *sand*, and *mud* or *silt* (Fig. 4.4). Combinations of two or more types are also common, for example, bedrock and gravel, bedrock and rubble, and sand and gravel.

4.3 LIFE IN RIVERS AND STREAMS

You are part of a community. If you are a town dweller, you know how different your community is from that in which your country friend lives. These two communities depend on quite different sources for food, livelihood, and entertainment. Their members differ in their ability to perform physical work and to sleep with a background of traffic noise. The niches available to members of the two communities are, on the whole, dissimilar. Humans adapt to the environmental conditions to which they are exposed.

Stream organisms are much the same. Only certain species inhabit given areas of a stream. Each has developed specialized features which enable it to survive in its particular environment. For example, you will remember that in Section 3.3 you met three types of mayfly nymph. None of these is adapted for life in a fast moving stream. Any mayfly nymph which you find in a swift current must have special adaptations to remain there. Look for these very carefully during your stream studies.

This section of the book is subdivided into four parts denoting four types of stream beds: *bedrock*, *rubble and gravel*, *sand*, and *silt and mud*. The common organisms of each will be briefly described and illustrated. These descriptions should help you identify the organisms during field trips. Note that we say *briefly* described. It is up to you to provide in-depth descriptions. During your field trips, you will find some animals that you have never seen before. Most stream organisms are fascinating creatures to study. Take one home with you. Study its habits and structure, and read about it in one of the *Recommended Readings*. Do this with such animals as insects and insect larvae, **not** with larger animals that are less frequently found and less readily replaced, for example, snakes.

As we discuss the organisms associated with the four types of stream beds, we will place each organism into one of three categories, according to its location in the stream. Bear in mind that some species may be found in more than one environment, but if you look closely, you will find some unique characteristic which makes each animal distinct. Thus a midge larva which lives in swift currents may be just a little different from one which inhabits slow sluggish waters.

(a) Bottom organisms are located in, on, or in close association with the stream bed. Since they are not as conspicuous as some other animals, you will have to search very carefully for these little beasts. Turn over stones and stir up the mud and vegetation on the bottom. Because the stream bed serves as a place for attachment, most of the organisms present in a fast moving stream will be of this type.

(b) Pelagic organisms are those which float or swim freely throughout the depth of the stream. Microscopic pelagic organisms make up the *plankton* which you met in Sections 3.4 and 3.5. You will remember that *phytoplankton* consists mainly of diatoms and desmids; protozoans, rotifers, and crustaceans are included in the *zooplankton*. The planktonic life is the aquatic "food crop," consisting of producers and consumers which provide food for higher order consumers. Few planktonic organisms live in rapid sections of streams, since they are swept downstream by the current. In these sections, the higher order consumers must rely on bottom plants and animals for their food supply. Planktonic organisms are abundant, though, in the slower waters of mud- and silt-bottomed streams.

(c) Surface organisms. Many aquatic insects such as water striders and water beetles, floating plants such as duckweed, and animals closely connected with or dependent upon these plants are all classified as surface organisms. Also included in this group are certain adult insects which live close to the stream and fly close enough to the water's surface to provide food for fish and frogs. For this reason, some adult insects will be illustrated along with their nymph or larval forms.

TYPES OF STREAMS AND ASSOCIATED LIFE

(a) Bedrock stream beds provide very little food and protection and, consequently, contain little life. One genus of blue-green alga, *Rivularia*, may be found growing in brown gelatinous masses on the rock (Fig. 4.5). The only other plant which can survive in this environment is a type of moss called fountain moss (Fig. 4.6). It appears as a dark green carpet. Irregularly branched, three-inch stems bear long, narrow leaves on their upper portions.

With this very limited supply of food, practically no animal life will be present. Nematodes (Fig. 3.9), which live in almost any environment, can be found if you search for them. These tiny, smooth, transparent worms move about by lashing their bodies in a whip-like fashion. They feed on any available plant material. If you view these little thread-like animals under a microscope you can distinguish the mouth end, which is blunt, from the sharply pointed tail end.

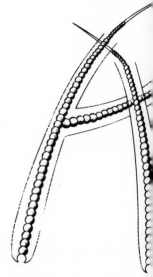

Fig. 4.5
The blue-green alga, *Rivularia*.

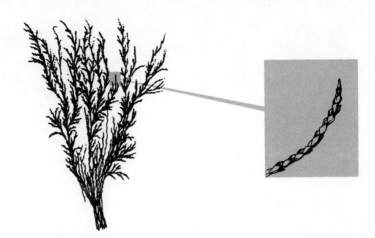

Fig. 4.6
Mountain moss, a stream plant which lacks roots.

Fig. 4.7
Mayfly nymph. Note the flattened body parts and gill covers.

Another animal which is able to survive in these streams is the mayfly nymph (Fig. 4.7). Obviously, though, it must be different from the three species observed in the pond (Fig. 3.19). This nymph is quite flattened and has two tail filaments rather than the usual three. In addition, it has many adaptations for adhering to the smooth rocks. These include a body with downcurved edges, thin blade-like legs, and very strong claws. Most important, a suction disc, formed by its plate-like gills, acts as an efficient holdfast.

You will find other species in a bedrock stream, since the type of vegetation on the banks, the source of the water, and the time of year, are also variables that affect the life in a stream.

There are many types of bedrock in various parts of the country. If you find a stream with a bedrock bottom, determine the type of rock and, perhaps, compare the life in this particular stream to that in another bedrock stream.

(b) Streams with rubble or gravel bottoms have a high stream velocity. They carry an ample supply of food and oxygen to any animals lying in wait among the nooks and crannies provided by the stony bed.

The "food crop" or "grass" of these streams is largely blue-green algae and fountain moss, as well as some species of diatoms (Fig. 4.8). These minute "pill-box" plants are most noticeable in spring, when they reproduce and cover the stones of the

Fig. 4.8
Some types of diatoms found in streams.

Pinnularia

Cymbella

Navicula

Amphora

Achnanthes

Life in Rivers and Streams   75

Fig. 4.9
The water hypnum, show
ing the leaf arrangement

stream bed with a blanket of golden brown. Many small primary consumers "graze" on these species. In addition, one species of water hypnum (Fig. 4.9) is present in almost any mountain brook or fast-moving stream. Note again the thin 3–4 inch stems which will not break off in the current. You can see bright green or yellowish masses of these plants forming a film over the stones in the bed. The water weed, *Elodea* (Fig. 3.35), is a larger plant which grows completely submerged, rooted in the gravel. It survives only in the slower streams of this type, since it has a rather weak root system. The dark green, translucent leaves are arranged in whorls of three along the brittle stems. A new plant can form from each fragment of a broken stem. With this tremendous rate of reproduction, *Elodea* often slows down and blocks up the water. It can thereby transform the stream into a slow muddy one. Snails, fish, and some crustaceans feed on this water weed.

Animal life is abundant if you search for it. You will undoubtedly find some fascinating little creatures. First, think about the bottom animals which hide under the rocks and stones. Collect one or two jars of stream water, being careful to include bottom plants, twigs, and decaying matter. Now examine the sample with the naked eye. Can you detect any movement? Yes? You should be able to trap some of these mysterious particles in an eye dropper. Transfer them to a microscope slide for examination. What do you see now? The tiny transparent lashing worms are nematodes. Can you distinguish another type of worm? The bristleworm, an oligochaete, is a small segmented worm with many tiny bristles. It is very closely related to the earthworm. In the muddy streams discussed later, it actually feeds in the same way, overturning the mud as an earthworm overturns soil. The most common genus on gravel bottoms is *Nais* (Fig. 4.10). It is usually found closely associated with the bottom plants and detritus. These tiny worms reproduce by *budding*, a process which is also used by *Hydra* (Fig. 3.8). Plan-

Fig. 4.10
The oligochaete, *Nais*.

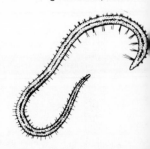

g. 4.11
wo common stream pro-
zoa. How are they
lapted to a life in flowing
ater?

Vorticella

Carchesium

aria (Fig. 3.11) are often found crawling over stones and bottom debris. Their flat bodies cling to the rocks to resist the current. Planaria are carnivorous and can easily be kept in captivity if fed any type of raw meat. Many interesting experiments can be performed on them. For example, if you cut a planarian longitudinally, each half will develop into a complete body. This and other experiments with planaria are described in most biology laboratory manuals.

If you have examined pond water under a microscope, you know what an amazing world exists there, a world inhabited by the protozoans. Although there are fewer types of protozoa in stream water than in pond water, there are still some intriguing individuals. *Vorticella* (Fig. 4.11) is a touch-sensitive protozoan with spring-like movements. It can be observed by scanning the surfaces of twigs, plants, and even many insects. Supported by a holdfast at one end, the "bell" of the animal waves around at the tip of a long fragile stalk. Thousands of cilia wave as *Vorticella* waits for food to come within reach. At the slightest touch, the cilia fold in and the stalk coils up like a spring. A few seconds later, it unfolds again. Similar to *Vorticella*, but a colonial species, is *Carchesium* (Fig. 4.11). These animals, which seem to be a mass of tiny bells on one stem, adhere to the undersides of sticks and stones. They grow as high as an inch or so. These are just two examples of the many types of protozoans you may find in the stream. You can spend hours watching and identifying others.

Now, turn over several of the larger stones in the stream. On the underside you will find a coating quite similar to that found on the top. Instead of plant growth, however, the coating consists of plant-like animals. Sponges and bryozoans (Fig. 3.6) appear very often on the same rock. A sponge is a skeleton of silica surrounded by a mass of living material. Flagella are present around the many pores and channels throughout the lattice. By their waving action they aid the flow of water in and out, carrying food and oxygen to the animal. Bryozoans, which are similar in appearance to moss, are made up of hundreds of thousands of individual animals living in a single colony. When you look at these masses, you see only the skeletons in which the animals are completely hidden. When undisturbed, the animals partially emerge from the skeleton, showing the tentacles which surround the central mouth. Each tentacle is covered with cilia to aid in directing food particles towards the mouth.

Among the stones on the bottom of the stream you will find a variety of insect larvae and nymphs. A very versatile little animal that can survive even in the most violent rapids and waterfalls is the midge larva (Fig. 3.10). You may easily mistake it for one of the nematodes or oligochaetes, but on close examination you will notice its distinguishing characteristics. It possesses true legs and appendages, mouthparts, and other insect characteristics. One genus, *Blepharocera*, the net-veined midge, has very powerful suction discs on its underside (Fig. 4.12). It can adhere to rocks even in the swiftest of rapids. You will see other midge larvae later which have adapted in another way—building cases in the muddy bottoms of slower streams. Adult midges resemble mosquitoes, but distinct differences can be seen when the animals are carefully examined. The wings of the mosquito are covered with scales while those of the midge are not. While resting, the midge holds its front feet off the surface. The mosquito, on the other hand, holds its hind feet up. Also the antennae of the midge are quite feathery; those of the mosquito are not.

Another well-adapted larva is that of the blackfly (Fig. 4.13). These larvae are most abundant in the northernmost cold streams and rapids, where they sometimes form a thick greenish covering on rocks and stones. The larva has a suction disc at the hind end of the body. It gathers food, mainly diatoms, by means of brushes at the mouth end. Since it is able to spin a silky thread from its salivary glands, it can move from rock to rock and travel upstream against the current without being swept away. The adult blackfly (Fig. 4.13) is a common pest. If you are a camper or fisherman, you are well acquainted with him. He is a tiny, black, hump-backed fly with very short antennae, and is thirsty for blood. Have you ever fed one?

Three insects are usually found in close association with one another in stony streams. They are the mayfly and stonefly nymphs and the caddisfly larva. You should now be quite familiar with the mayfly nymph. Some of these, which have extremely flattened bodies, adhere to the undersides of stones, while others construct burrows for protection. They are swift, agile, and very adept at catching food particles that flow past them. For this purpose they have long front "claws" which are, in reality, specialized front legs (Fig. 4.14). The nymph has feathery gills either on the back or side of the abdomen. This is the best position for obtaining the maximum amount of oxygen which these active insects require. The adult mayfly is a graceful insect which, when at rest, holds its front legs and its wings in an upright position. The nymph feeds mainly on diatoms while the short-lived adult does not feed. Adult mayflies fly low over the streams and are a favorite food of many fish. For trout fishing, you may have used manufactured flies or lures which are patterned after mayflies.

Stonefly nymphs (Fig. 4.15) can easily be mistaken for those of mayflies if you do not know the differences. First, very few

Fig. 4.12
The larva of the net-vein midge, *Blepharocera*. dorsal side, (B) vent side showing sucti discs.

Fig. 4.13
The blackfly. (A) larva, a (B) adult.

. 4.14
nayfly nymph found in
ny streams. Note the
g front "claws."

. 4.15
e stonefly. (A) nymph,
d (B) adult.

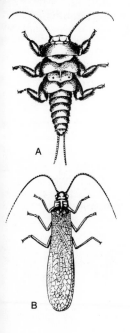

A

B

mayfly nymphs have only two tail filaments; most of them possess three. All species of stoneflies have only two filaments. Second, the stonefly nymph has extremely long antennae. The antennae of the mayfly nymph are short. Further, stonefly nymphs have two claws on each foot, mayfly nymphs only one. The gills of stonefly nymphs are on the ventral side, usually at the base of each leg, while on the mayfly nymph they are dorsal or lateral. Because they do not like bright light, stonefly nymphs are always found clinging to the undersides of rocks with their strong claws. If you overturn their rock, they scatter madly all over its face until they drop off the edges into the water. Their flattened bodies allow them to move through the current and to squeeze between and under rocks and gravel. The stonefly nymph spends over a year in the water. Then it sheds its nymphal skin for the last time and becomes an adult insect. The final nymphal stage of the mayfly is spent on land, which further differentiates the two insects.

The adult stonefly (Fig. 4.15), although similar to the mayfly in flight, is not nearly as dainty and beautiful when viewed more closely. It is dull brown or yellow, squarely built, and holds its wings down over the abdomen. Notice the same long antennae and double claws as on the nymph. In some species the adults emerge in January or February and can be seen running about over the snow-covered banks of streams. Those which emerge in spring and summer spend much of their time hiding on the undersides of leaves of trees which overhang the stream. They very rarely use their wings. If the wind shakes the branches many of these flies fall to the ground or into the water. On the ground they try to run and hide as quickly as possible, but in the water they are helpless prey for hungry fish.

Caddisflies, easily mistaken for small moths, have been made famous by the larva which is very clever in the art of home construction. You may often see little piles of sticks, stones, or other debris moving slowly along the stream bed and not think twice about it. But, if you pick up this little case, you will find the caddisfly larva, or caddis-worm, tucked neatly inside (Fig. 3.20). The larva of one genus, *Rhyacophila* (Fig. 4.16), does not build a

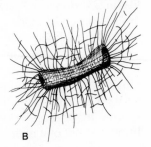

Fig. 4.16
Caddisfly larvae. (A) *Rhyacophila*, and (B) *Polycentropus*, showing the structure of its cleverly built net.

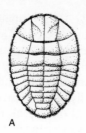

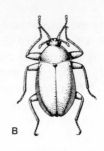

A   B

Fig. 4.17
The riffle beetle. (A) lar
commonly known as t
"water penny," and (
adult.

Fig. 4.18
The damselfly nymph
the genus *Agrion*. Note t
adaptations for life in fa
moving streams.

case until it is ready to pupate. Instead, like mayfly and stonefly nymphs, it crawls under stones in riffles until it is fully grown. Other species build intricate cases of all shapes and sizes, as Figure 3.20 shows. Some species cement one end of the case to rocks or aquatic plants, and partly emerge from the free end to capture food. Others crawl about using their strong claws, dragging their cases with them until it is time to pupate. Then their cases, too, become cemented to the rocks. The most fascinating caddis-worms are those which build or spin nets to aid in catching food. *Poly-centropus* (Fig. 4.16) builds a beautiful web-like net which is attached at one end and free at the other. It suspends the net longitudinally in the stream, waiting for the current to wash food in. The caddis-worm *Macronema* lies naked in a trap which it builds from small stones. This cup-shaped trap opens upstream, allowing the current to wash food in to the waiting larva.

A close neighbor of the mayfly nymph, stonefly nymph, and caddisfly larva is the larva of the riffle-beetle, commonly called the "water penny" (Fig. 4.17). These oval, brownish larvae cling tightly to rocks and stones of rapids and fast streams. When you view them from above, you can easily overlook them. But, if they become dislodged from the rock, you will quickly see the flashy white gills and waving legs. The adult riffle beetle (Fig. 4.17) never strays far from the water. It is usually found resting on stones which protrude from shallow streams.

A fierce predator in streams as well as ponds is the dragonfly nymph (Fig. 3.22). Dragonfly nymphs commonly inhabit slower muddy streams. A few species of damselfly nymphs live in fast streams. They are quite similar to dragonfly nymphs but are more slender. One genus, *Agrion* (Fig. 4.18), is well adapted, with a streamlined shape, sloping head, and upturned gills to take oxygen from the passing water.

Two closely related insects, whose larvae take cover in gravel and sandy streams, are the alderfly and dobsonfly (Fig. 4.19). The latter preys on the former; thus you will not find the two species in close proximity to each other. The alderfly larva is brown, about one inch in length, and has strong jaws with which it

Fig. 4.19
The alderfly (A), and d
sonfly (B), two closely
lated insects of the strea
community.

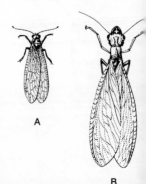

A   B

catches caddis-worms. The adults are clumsy fliers and rely on their legs as a mode of escape. The dobsonfly, both larva and adult, is very similar to its relative, except that it is larger. It grows to a length of two to three inches and is a fierce-looking animal.

It is unusual to find an aquatic caterpillar. The larvae of at least one species of moth, however, inhabit swift streams (Fig. 4.20). Often this caterpillar is found on the same rocks as mayfly and stonefly nymphs. Although it cannot live completely submerged, it breathes and eats in the water and must be constantly washed by waves. It builds itself a silken home which is cemented to the rock. It leaves a few openings, of course, through which the water flows. The adult emerges from the pupa in summer and stays quite close to the stream. Its white hind wings flash brightly behind the dull brown fore wings.

Fig. 4.20
An aquatic moth larva.

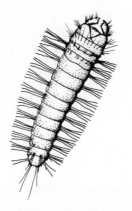

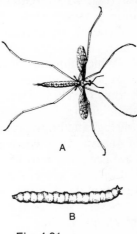

A

B

Fig. 4.21
The cranefly. (A) adult, and (B) larva.

A terrifying looking insect in the adult form is the cranefly (Fig. 4.21), a relative of the mosquito. It is often called the "mosquito hawk," despite the fact that it lives entirely on nectar from flowers. The larvae are both carnivorous and herbivorous. The genus most commonly found in rapid streams is *Antocha*. Look for a silky case, 1 or 2 inches long, wedged in a crevice or between two stones. Within this little case is the cranefly larva (Fig. 4.21).

The water strider, a surface organism discussed in Section 3.1, is commonly mistaken for a spider. It has only six legs, however, and thus is a true insect. The genus found in rapidly moving water is *Rhagovelia*. It has a specialized oar-like structure composed of tiny hairs on its middle legs. This structure acts as a rudder when *Rhagovelia* wants to travel upstream.

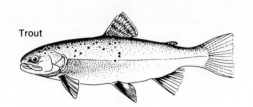

Trout

Fig. 4.22
Some of the fish co
monly found in fast-m
ing streams.

Other animals which are found in streams as well as in ponds are snails, clams, leeches, and amphipods. All of these creatures are well-adapted to life in streams, having suction devices, streamlined shapes, or strong appendages. How is the amphipod (Fig. 3.15) adapted to this way of life?

Who is better adapted for life in moving water than the crayfish (Fig. 3.13)? With his strong walking legs, powerful "tail," and huge grasping claws, he can withstand almost any current. And, of course, an abundance of food is available for him in the stony bed.

Which animals have we not mentioned yet? The largest and most popular of the stream organisms—fish (Fig. 4.22). These are the only vertebrates mentioned in this section. Trout, of course, are abundant in fast streams. They hide in the shade of overhanging logs and branches, waiting for a fly to fall into the water or for an unsuspecting worm or aquatic insect to pass by. Speckled trout are generally 6 to 15 inches in length and are unmistakably marked with tiny colorful dots. Other species of trout are present in many parts of the country.

Several species of minnows can be caught using a seine net. The black-nosed dace is less than 3 inches long. It hides among stones and gravel, and can be identified by the black band down each side of the body. The horned dace is a dark bluish fish with a black spot at the base of its dorsal fin. It often reaches a length of 10 inches. An easily located minnow is the flashy shiner which rarely grows larger than 5 inches. In breeding season, the male develops tiny horns on its head, distinguishing it from the female. Another minnow common to gravel bottom streams is the bluntnosed minnow. It is olive-colored with bluish sides and about 2–4 inches long. One of the smaller minnows is the stickleback, which is easily identified, especially if you grasp it in your hand! Its sharp dorsal spikes protect it from its enemies. The stickleback, a mere 2½ inches long, is an excellent fish to keep in your aquarium. Its antics and mating dances are fascinating to observe. Darters, like sticklebacks, are very small, seldom exceeding 2½ inches. They are very thin fish and are quite distinctive because of their bright coloring. The lightning-quick antics they perform are similar to those of the rabbit in flight.

Horned dace

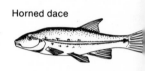

Shiner

Black-nosed dace

Blunt-nosed minnow

Stickleback

Rainbow darter

Johnny darter

Only the most common types of fish have been mentioned here. You should use a fish key to help identify species.

(c) Sandy streams are not common, at least not in long stretches. This is the least productive type of stream. There is no solid rooting material for higher plants and no large smooth surfaces for the attachment of algae, mosses, and other plants. Any food for consumers must be either washed down from upstream or grown along the margins. Some of the animals which can survive in these conditions are planaria, nematodes, mayflies, caddisflies, and alderflies. All of these have been described earlier.

Of course, streams of different types can contain some sand. In these cases, animals such as those above will inhabit the sandy areas.

(d) Muddy or silt-bottomed streams are generally quite high in productivity. As more and more water flows through this type of stream, a buildup of mud and silt occurs. The stream gradually takes on the characteristics of ponds. Since there is abundant rooting material, higher plants can grow and thrive. Besides diatoms, algae, fountain mosses, and hornworts, plants such as pickerel weed, water weed, burreed, arrowhead, and watercress appear in the shallow regions at the edges of the stream. For descriptions of these plants, see Section 3.8. In extremely slow waters which are becoming stagnant, duckweed (Fig. 3.40) may begin to grow and eventually cover the water's surface.

With this large number of producers, the number of consumers also rises greatly. Rotifers (Fig. 3.26) and copepods (Fig. 3.27) are two little animals which, until now, have only appeared in ponds. The rotifers in streams are attached, usually to some plant material. The wheel-like organs at the anterior end catch food particles which are drifting by. Copepods, a type of crustacean, appear as tiny specks jerking their way through the water. The most common genus, *Cyclops*, is not found in streams as often as *Diaptomus* and *Canthocamptus*.

Protozoans are generally abundant in these slower waters. You will likely find most of the types common to faster streams and to ponds, along with many others. Don't forget to take a sample of the water, mud, and bottom debris back to the laboratory with you.

The worms that you find in the mud will be quite numerous. Nematodes will abound, burrowing in the mud and silt. Two oligochaetes, *Tubifex* (Fig. 3.9) and the bristle-worm (Fig. 4.10), are sure to exist wherever there is mud. These two oligochaetes seem to thrive in oxygen-deficient conditions. As their numbers increase, you can be sure that the oxygen level is falling.

The bryozoans which we met in faster streams reappear, attached to stones, logs, and aquatic plants.

Mayfly nymphs, dragonfly nymphs, and caddis-worms again appear. As you might expect, they have unique adaptations for this environment. In muddy or silty surroundings, the gills must be protected so that they are not clogged or covered with debris. Figure 4.3 shows the covering over the gills of the mayfly nymph, *Caenis*. You will also find here the burrowing and sprawling types of mayfly nymphs, which occur in ponds.

Look back at the damselfly nymph in Figure 4.18. Note the long gill filaments at the end of the body. This type of structure would not work in muddy water. The gills of the dragonfly nymph, however, are contained in a rectal sac that protects them from mud or silt (see Figure 3.22.). The midge larva (bloodworm) in muddy streams is a burrower and looks much like a nematode or an oligochaete (Fig. 3.10). The amphipod, crayfish, and leech inhabit these waters with very few adaptive changes since they already are equipped to live in many types of surroundings. How?

The large water strider (Fig. 3.1) is replaced in slower streams by one which is much smaller. (But note, the larger one may still appear.) *Gerris marginatus* is about ½ -inch in length and is winged (Fig. 4.23). Another aquatic insect found in sluggish streams is the water boatman (Fig. 3.33). The specialized hind legs are flattened for swimming and diving. These animals spend a large part of their time in the plant growth on the muddy bottom. Several species of snails and clams, generally the same as those found in ponds, also live in the mud.

The fish found in slow streams (Fig. 4.24) are quite different from those found in fast streams. They do not have the powerful muscles needed by fish which inhabit faster streams and must

Fig. 4.23
The water strider, *Ge marginatus*.

Fig. 4.24
Some fish found in mu streams.

Common
sucker

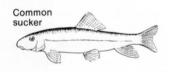

Yellow perch

Northern pike

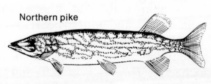

Catfish

Miller's thumb

combat strong currents. The common sucker is easily identified by its round sucking mouth which is ventral on the head. These fish may become 12–13 inches in length. They are quite sluggish and easy to catch. The catfish is very familiar to anyone who has fished in ponds or streams. Its distinguishing characteristics allow no mistakes in identification. The northern pike is another inhabitant of sluggish streams. Here it hides among the weeds, now and then lurching forward after another fish or a frog. The largest pike are about 24 inches long but stream pike generally do not attain this length. Its identifying characteristic is the pointed head with its protruding lower jaw. The yellow perch is a lake and pond fish which may be found in sluggish streams. Its length is between 6 and 10 inches. It has an olive back that blends with yellow sides marked by vertical black stripes. Why do you suppose it has these stripes? Lastly, the "Miller's thumb" is quite a comical looking little fish. His dull, olive-green body is mottled with spots of dark brown, camouflaging him against the muddy background. The 5-inch long body is quite smooth except for a warty head.

Only a few of the more common fish have been described. You may or may not find these species in your area. But, you *will* find many different types no matter where you perform your field studies.

As you step from the stream onto the bank another world appears. It is strongly dependent upon the one which we have just described. Do not forget the trees, bushes, frogs, salamanders, snakes, turtles, muskrats, raccoons, water birds, and so on. Many of these are probably more familiar to you than most of the aquatic organisms that we have described here. We suggest that you identify as many of them as you can, study their habits and habitats, and try to determine the niches that they occupy in the community that you are studying.

For Thought and Research

1 (a) Why is a muddy or silt-bottomed stream generally high in productivity?

(b) Why does a stream with a rubble or gravel bottom generally have more types of living organisms in it than a stream with a bedrock bottom?

(c) Why are streams with sandy bottoms the least productive type of stream?

2 (a) Make a list of the aquatic organisms discussed in this unit that are streamlined. In what kind of stream environment are these organisms usually found? Why?

(b) Why are most zooplankton not streamlined?

3 Rule a page so that it has four columns. Place the following titles on these columns: BEDROCK; RUBBLE OR GRAVEL; SAND; MUD OR SILT. Under each of the titles, make a list of the stream organisms that may be found in that environment. This summary will be useful to you during field studies.

4 You can discover many of the characteristics of streams in the laboratory by performing Investigations 12-6 and 12-7 in *Investigating the Earth*, Earth Science Curriculum Project, Houghton Mifflin Co., 1967. Perform these investigations to discover:

(a) how the size and shape of rocks affects the way they are transported in streams;

(b) the relationship between stream slope and rate of erosion;

(c) the relationship between stream volume and rate of erosion;

(d) the effects of erosion.

5 Select an insect larva or nymph that commonly inhabits only streams or rivers. Carry out the studies suggested in question 8 of the *For Thought and Research* section of Unit 3.

6 Set up a classroom aquarium containing stream minnows and stream plants that you collect on a field trip. Minnows 1–2 inches in length have a much higher survival rate than larger minnows. For information regarding the maintenance of an aquarium, consult *A Source Book for the Biological Sciences* by E. Morholt et al., Harcourt Brace Jovanovich, 1968. Pay particular attention to the ways in which the fish have adapted to the environmental conditions in which you found them. Return the fish to a stream after you have completed your observations.

7 Complete the laboratory studies listed in question 11 of *For Thought and Research* in Unit 3.

Recommended Readings

1 *The Life of Rivers and Streams* by R. L. Unsinger, McGraw-Hill, 1967. This easy-to-read book contains interesting descriptions and color photographs of the ecology of rivers and streams.

2 Consult *Recommended Readings* 4–13 of Unit 3 for the identification of organisms that are not discussed in this book.

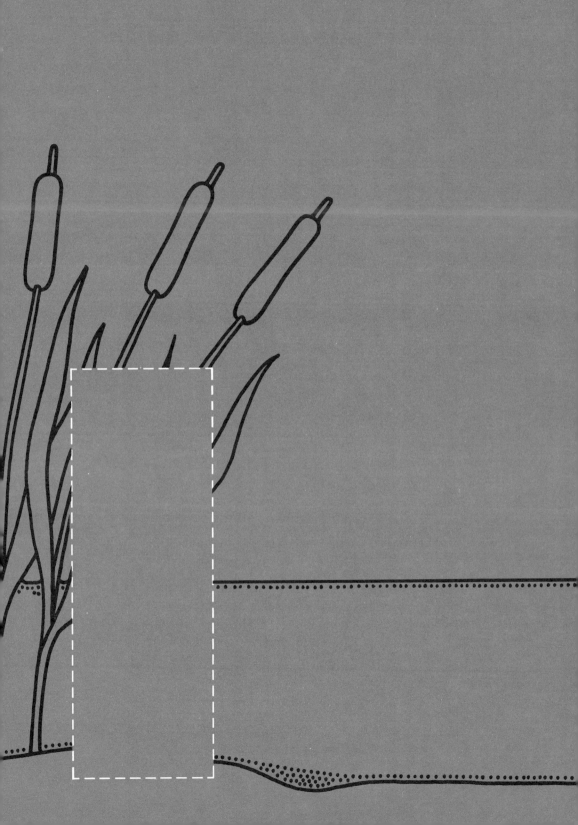

Field and Laboratory Studies

5

In this Unit, studies of the chemical, physical, and biological characteristics of waters are outlined. Some are simple, some are not. It is important to remember, however, that each factor, no matter how trivial it may seem, contributes to the character of the body of water. Each study must be carried out as if its subject is the most important feature of the water.

Some of these studies are best done when you have reached a definite place in your reading. Where this is so, the appropriate studies are listed in the *For Thought and Research* portion of the section concerned. Many of the studies, however, cannot be done until you go on a field trip. These include such things as the measurement of stream velocity and the determination of a pond profile. Still others give useless results unless they are performed in the field in conjunction with other studies. The D.O. test is an example of such a study. Nevertheless, you should still rehearse a test like this before you go on a field trip. Try it on the water in a classroom aquarium. It is important that you get to know the equipment and the procedures for such a test. Then, once you are in the field, you can perform the test quickly and accurately.

It is not intended that you do everything that is in this Unit. Whether or not you do some of these studies depends on the equipment available and on the type of study that you wish to perform.

Also, the sequence in which you perform these studies will vary from the sequence here. In general, studies of physical factors have been grouped together, as have studies of chemical factors and of biological factors.

5.1   DISSOLVED OXYGEN (D.O.)

The test for dissolved oxygen (D.O.) is probably the most important test for determining water quality. The best and easiest methods for D.O. analysis use D.O. testing kits (such as the Hach or LaMotte) or a D.O. meter. The kits use pre-weighed and pre-measured amounts of chemicals which are added to a water sample in a certain order. Titration is done on a direct-count basis or with a micro-buret. Oxygen meters are generally more accurate than kits, but are considerably more expensive and are temperamental.

Most kits are based on some modification of the Winkler Method. If you do not have kits, you can use the modified Winkler test below. It is accurate to $\pm$ 0.5 ppm.

Materials

a)   250–300 ml bottles with ground glass stoppers (B.O.D. bottles)

b)   500 ml and 1 liter Erlenmeyer flasks

c)   100 ml volumetric pipette

d)   2 10 ml Mohr pipettes

e)   buret

f)   manganese (II) sulfate solution: 480 gm of $MnSO_4 \cdot 4H_2O$, 400 gm of $MnSO_4 \cdot 2H_2O$, or 364 gm of $MnSO_4 \cdot H_2O$ dissolved in enough distilled water to make 1 liter of solution.

g)   alkaline-iodide solution: 500 gm of sodium hydroxide (or 700 gm of potassium hydroxide) and 150 gm of potassium iodide (or 135 gm of sodium iodide) dissolved in enough distilled water to make 1 liter of solution.

h)   sulfuric acid: concentrated.

i)   starch solution: 5 gm of soluble starch dissolved in 1 liter of distilled water. Sterilize the solution and store it in small bottles, to be opened as they are needed. This prevents rapid deterioration of the solution.

j)   sodium thiosulfate solution: dissolve 3.953 gm of $Na_2S_2O_3$ in enough distilled water to make 1 liter of 0.025M solution. Add 5 ml of chloroform to preserve it. A new solution should be made up every 3 to 4 weeks. This solution must be standardized occasionally with 0.025M potassium dichromate as described below.

Standardizing the Sodium Thiosulfate Solution

a)   Dissolve 2.5 gm of KI in 50 ml of distilled water.

b)   Add 0.5 ml of concentrated sulfuric acid.

c)   Add 20 ml of 0.025M potassium dichromate solution (6.96 gm per liter of solution).

d)   Place the solution in a dark place for 5 minutes.

e)   Titrate this solution with the sodium thiosulfate solution to be standardized, using a few drops of starch solution as the indicator.

If the sodium thiosulfate solution is 0.025M, only 20 ml are required to reach the end point. To find the correction values, use the following rules:

If more than 20 ml were required, then the solution is weaker than 0.025M and the correction value should be less than 1. For example, if 21 ml were used, divide 20 by 21 to give 0.95, the correction factor. Thus, if 13 ml of sodium thiosulfate solution are used in a D.O. titration, the corrected D.O. value (ppm) is 13 times 0.95, or 12.4 ppm.

If less than 20 ml were required, the solution is stronger than 0.025M and the correction value is more than 1. For example if 17 ml were used, divide 20 by 17 to give 1.18, the correction value. If 8 ml of the sodium thiosulfate are used in a D.O. titration, the corrected D.O. value in ppm is 8 times 1.18, or 9.4 ppm.

Procedure

a)   Completely fill a 250–300 ml sample bottle (preferably a B.O.D. bottle) with the water to be tested. Be sure no air enters the water during this process.

b)   Add 2 ml of manganese (II) sulfate solution and 2 ml of alkaline-iodide solution, *beneath the liquid surface*. Use a separate pipette for each solution. Caution: Do not mix the water, or contamination with the atmosphere will result. Be sure that the excess solution overflows onto a surface that cannot be harmed by it.

c)   Replace the stopper, making sure that no air bubbles are trapped beneath it. Shake the mixture vigorously and then allow the precipitate to settle to the lower half of the bottle.

d) Add 2 ml of concentrated sulfuric acid above the water level; replace the stopper carefully. Shake. At this point titration can be delayed several hours without affecting the result.

e) Transfer 200 ml to a 500 ml Erlenmeyer flask using a volumetric pipette.

f) Titrate with $0.025M$ sodium thiosulfate solution. One or 2 drops of the starch solution should be added when the color of the solution has become light yellow (after the addition of some of the sodium thiosulfate). Continue titrating until the blue color is gone.

g) The D.O. in ppm is the number of ml of $0.025M$ sodium thiosulfate solution used to reach the end-point, subject to correction as noted above.

Notes

a) All steps before the final titration must be done in the field immediately upon collection of the water sample. The final titration can be done a few hours later.

b) Caution must be used in handling these chemicals.

c) All of the required solutions can be purchased, pre-standardized, from most chemical supply companies.

d) If you have studied chemistry, you should look up the reactions that are involved in this D.O. determination.

5.2 FREE CARBON DIOXIDE

The carbon dioxide content of waters is important from an economic viewpoint, since this gas contributes to several forms of corrosion. Biologically, a concentration of carbon dioxide greater than 25 ppm can be lethal to aquatic animals. Further, a high carbon dioxide concentration is usually accompanied by a low D.O. concentration.

Test kits of the Hach and LaMotte type are available for measuring the carbon dioxide content of water. The following procedures can also be used.

Materials

a) several 100 ml, short form (33 x 200 mm) Nessler tubes

b) droppers

c) buret

d) phenolphthalein indicator solution

e) 0.0227M sodium carbonate solution

Procedure

a) Collect 100 ml of the sample water in a Nessler tube. Avoid agitation and contact with the air. Proceed as quickly as possible with the following steps in this procedure.

b) Add 10 drops of the phenolphthalein indicator. If a pink color forms, no carbon dioxide is present.

c) If the sample stays colorless, titrate with the 0.0227M sodium carbonate solution until a faint but permanent pink color forms. Do not agitate the sample during the titration, but rotate the tube gently in order to mix. This avoids contamination from atmospheric carbon dioxide.

d) The free carbon dioxide present in the water, in ppm, is the number of ml of sodium carbonate solution used multiplied by 10.

5.3 *pH*

pH can be measured by several methods. One easy and fairly reliable method uses a *pH* kit with a universal indicator and a color comparator (Fig. 5.1). The accuracy of such a kit is usually limited to ± 0.5 *pH* units. Just as convenient, reliable, and accurate are the Fisher Alkacid test papers, also shown in Figure 5.1. For higher accuracy, a portable *pH* meter can be used. Such devices are expensive though, and often difficult to maintain. Accuracy almost as high as that with a *pH* meter can be obtained using the following method:

Materials

a) 25–50 ml bottles

b) color charts or color discs for each indicator

Indicator	pH Range
Bromocresol purple	5.2–6.8
Bromothymol blue	6.0–7.6
Phenol red	6.8–8.8
Cresol red	7.2–8.8
Thymol blue	8.0–9.6

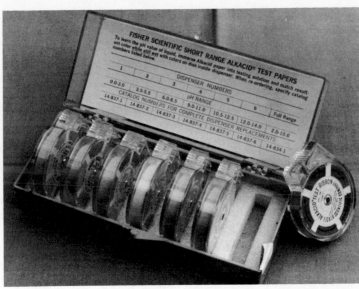

Procedure

a)   Place about 10 ml of the water to be tested in each of 5 bottles.

b)   To the first add 4 or 5 drops of bromocresol purple; to the second add 4 or 5 drops of bromothymol blue; and so on. Mix each sample by rotating the bottle. Do *not* place your thumb over the end and shake because this will contaminate the sample.

c)   Compare the colors obtained with the color charts or discs, and read the *p*H directly from them.

5.4   ALKALINITY

Tests for alkalinity are easily done with small kits (like Hach and LaMotte). However, if you wish to do your own chemistry, the test is described below.

Materials

a)   250 ml Erlenmeyer flask

b)   droppers

c)   buret

d)   100 ml volumetric pipette

e)   phenolphthalein indicator solution

f)   methyl orange indicator solution

g)   sulfuric acid—$0.01M$

Procedure

a)   Add 5 drops of the phenolphthalein indicator to a 100 ml sample of the water to be tested.

b)   If the solution becomes colored, proceed with step (c); if not, go directly on to step (d), skipping (c).

c)   If the solution became colored in step (b), titrate with the $0.01M$ sulfuric acid until the color disappears. Record the number of ml of acid used. To find the phenolphthalein alkalinity, in parts per million (ppm) as $CaCO_3$, multiply the number of ml of acid used by 10. (The phenolphthalein alkalinity measures the amount of hydroxide and half the amount of carbonate in the water.) Continue with step (d).

d)   Add 5 drops of methyl orange indicator to the solution from step (b) or (c). If the solution becomes yellow, titrate with the $0.01M$ sulfuric acid until a pinkish color appears and persists. Record the ml of acid used.

e)   The methyl orange or total alkalinity is the sum of the ml of acid required for both of the titrations, multiplied by 10. This gives the total alkalinity in ppm as $CaCO_3$. Total alkalinity includes hydroxide, carbonate, and bicarbonate.

Note: All titrations should be done over a white surface so that color changes can be spotted quickly.

5.5   TOTAL HARDNESS (T.H.)

Hardness can be measured best with any one of the kits of the Hach or LaMotte variety. You can even buy kits for this purpose from pet shops.

The soap method for determining hardness is described below. No attempt has been made to include ways of correcting for the lather factor, since in most cases an error of $\pm$ 10 ppm is acceptable. With several commercially available water test kits, errors can be as high as $\pm$ 16 ppm.

Materials

a)   250 ml glass-stoppered bottle

b)   50 ml pipette

c) droppers

d) 500 ml and 1 liter volumetric flasks

e) stock soap solution: shake 100 gm of pure castile soap powder into 1 liter of 80% ethyl alcohol. Cover and let stand for 1 to 3 days, then decant the upper layer. Discard the soap left in the bottom of the bottle.

f) standard soap solution: dilute a small portion of the stock soap solution with 80% ethyl alcohol until 1 ml is equivalent in hardness to 1 ml of standard calcium chloride solution.

g) standard calcium chloride solution: dissolve 0.5 gm of anhydrous calcium chloride in a few ml of dilute hydrochloric acid. Add 200 ml of carbon dioxide-free distilled water. Neutralize the solution with ammonium hydroxide until it is just alkaline (use litmus as an indicator). Add further distilled water until the volume of the solution is 500 ml. Store in a glass container. One ml of this solution is equivalent in hardness to 1 mg of calcium carbonate, which, in turn, is equivalent in hardness units to 1 ml of standard soap solution. Thus it is necessary for you to test the soap solution and adjust its concentration until 1 ml of the solution just forms a permanent froth when it is shaken with 1 ml of the standard calcium chloride solution.

Procedure

a) Place a 50 ml water sample into a 250 ml bottle.

b) Add standard soap solution in 0.2 ml drops, and shake the bottle after each drop.

c) When a lather is first seen, let the bottle stand for 5 minutes. If the lather remains, the end-point is reached. If the lather disappears, continue to add soap solution until the lather does remain.

d) The number of ml of soap solution used, multiplied by 20, gives the hardness (in terms of calcium carbonate) in ppm.

5.6 TOTAL SUSPENDED SOLIDS (T.S.S.)

Materials

a) fine filter paper

b) analytical balance

c)   1 liter bottles

d)   funnel

Procedure

a)   Weigh a filter paper.

b)   Filter a 1-liter sample of water through the weighed filter paper.

c)   Allow the filter paper to dry completely.

d)   Reweigh the filter paper. The change in weight is the weight of the total suspended solids (T.S.S.) in 1 liter of water. T.S.S. values are commonly expressed in ppm (mg per liter).

5.7   TOTAL DISSOLVED SOLIDS (T.D.S.)

The experiment for testing the T.D.S. is rather simple, but it is exactingly quantitative. Accurate tests cannot be easily done except in a laboratory equipped for them. The method below, however, will give you reasonably accurate results.

Materials

a)   250 ml beakers

b)   dustproof chamber (if available)

c)   analytical balance

d)   electrical hot-plates

e)   volumetric pipette

Procedure

a)   Weigh to the nearest 0.0001 gm, if possible, 3 clean, dry 250 ml beakers.

b)   Place 100 ml of the filtrate from the T.S.S. experiment (outlined in Section 5.6) in each of these beakers.

c)   Slowly and carefully evaporate to dryness using electrical hot-plates and a dust-proof chamber. Do not let the beakers get too hot or some of the dissolved solids may be vaporized or decomposed.

d)   Make the appropriate calculations for each of the 3 samples to determine the weight in grams of the solids dissolved in 100 ml of water. Average the 3 values that you obtain.

e)   Convert your answer to ppm by multiplying the value obtained in (d) by 10,000.

5.8   TURBIDITY

This is another method of estimating the amount of suspended matter in a water sample. Turbidity is measured with an instrument called a turbidimeter and relies on how much a beam of light is scattered. A turbidimeter is supplied in many larger water testing kits, but such kits are quite expensive. Even alone, a good turbidimeter is expensive. A less expensive one, the Jackson water turbidimeter, is sufficient for most purposes. The standard turbidity unit is the Jackson Turbidity Unit (J.T.U.). The Jackson water turbidimeter is reliable when the turbidity is above 25 J.T.U. The turbidity value of water that has been treated for drinking purposes should be less than 1 J.T.U. Untreated sewage or silt-laden river water can have a turbidity in excess of 1,000 J.T.U.

5.9   CONDUCTIVITY

Conductivity is linked closely to total dissolved solids (T.D.S.) of the water. Both are indicators of the productivity of waters, since they are measures of the nutrient material dissolved in the water. In general, high conductivity and high T.D.S. are associated with fertile lakes, and low conductivity and low T.D.S. with relatively infertile lakes.

      Conductivity can best be measured with a conductivity meter. This instrument measures the ability of the water to conduct an electrical current, which, in turn, depends on the number of ions in the water.

      In both conductivity and T.D.S. tests, the water sample must be allowed to settle or, preferably, be filtered in order to rid it of suspended solids.

      Before a conductivity test can be run, the temperature of the water sample must be taken. (The amount of material dissolved in water is dependent on the temperature of the water.) A standard temperature of 25°C is usually used. If the sample is warmer or colder, one of two things can be done. The temperature can slowly be brought to 25°C, or the test can be done at the different temperature. In the latter case a correction value must be used, to correct to the standard temperature. Charts with correction values on graphs are available from many chemical supply houses. Many conductivity meters have a built-in temperature compensator.

5.10 TRANSPARENCY

An indication of the amount of suspended matter in water can be obtained with a Secchi disc. This is a metal disc, 20 cm in diameter and divided into four quarters, 2 of which are white and 2 black (Fig. 5.2). Several kinds of information can be gathered with the disc: a rough measure of the suspended matter; the depth of reflected light penetration; and a rough estimate of the extent of the littoral zone.

Fig. 5.2
A Secchi disc.

To obtain a Secchi disc reading, lower the disc into the water, in the shade, until it just disappears. Take a depth measurement at this point. Then raise the disc until it just reappears. Take another depth measurement. The two measurements are averaged. This process is repeated three times and the overall average is considered to be the proper Secchi disc reading. The surface conditions and the color of the water should be noted when performing this test. By looking through a glass-bottomed bucket, much better measurements can be made since surface distortions are eliminated (Fig. 5.3).

If the Secchi disc reading is low (for example 5 feet), the water contains much suspended matter. If, on the other hand, it is very high (for example 30 feet) the water is quite clear and hence, relatively free of suspended matter.

It has been suggested that the Secchi disc reading corresponds closely to the depth limit of the littoral zone, the area of rooted plant growth. Thus, if the Secchi disc reading is 15 feet, the 15 foot contour of the lake is roughly the boundary of the littoral zone.

The Secchi disc reading, however, can vary from place to place, from hour to hour, and from day to day, depending on above-water conditions. When you are trying to determine the location of the littoral zone, test repeatedly over a period of a few months, in order to get the best approximation of the zone boundary.

5.11 COLOR

The color of the water gives an indication of the amounts of suspended and dissolved matter present in the water.

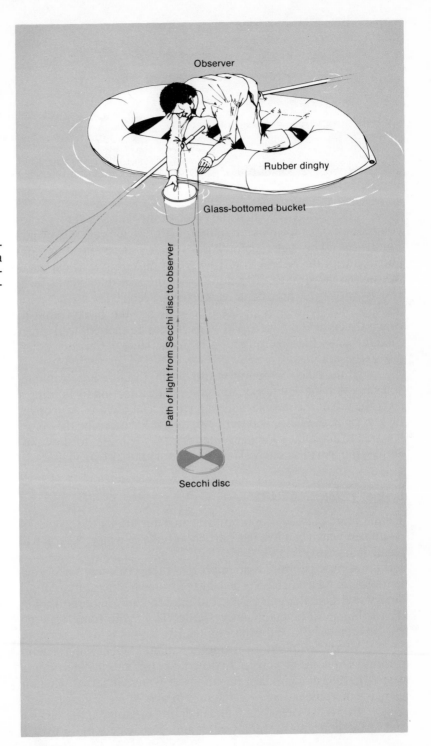

Observer

Rubber dinghy

Glass-bottomed bucket

Path of light from Secchi disc to observer

Secchi disc

g. 5.3
ie use of a glass-bot-
med bucket to obtain a
oper Secchi disc read-
g, free from water sur-
ce distortions.

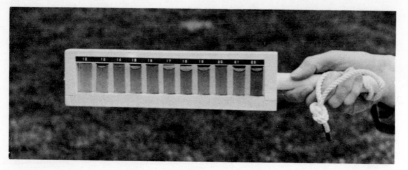

Fig. 5.4
Standard Forel-Ule Scal[e]
for water color determin[a]
tion.

The best way to find the color of the water is with a color comparator. One form of the standard Forel-Ule Scale is shown in Figure 5.4. The water color is most easily determined in conjunction with the Secchi disc. Lower the disc until it lies about 3 feet below the surface. The number of the vial that blends most closely with the water color against the Secchi disc is the color number. The whiteness of the disc provides the background to which the color is referred. The vials should be shaded from direct sunlight when the determination is made, since reflections could distort the readings.

If the color comparator and Secchi disc are not available, you can often make a reasonable color determination by holding a test tube full of the sample against a white background. Adjectives such as light, medium, and dark can be used to describe the color. It is best to allow the sample to stand for a few minutes to allow solids to settle. An inexpensive Hach color test kit is also available.

5.12   TEMPERATURE

Temperature plays an important role in determining the species of organisms which can live in a particular body of water. As you will recall, it also affects D.O. levels.

Air temperature is an important factor to know since it determines, to a large extent, the surface water temperature. An ordinary thermometer can be used to take the air temperature, but the reading must be taken in the shade. (Hold your hand between the sun and the bulb of the thermometer.)

Water temperature is another matter. Generally, a temperature *series* must be taken. In deep lakes this aids in determining where the thermocline lies; in shallower water, it shows whether there is a thermocline, or even the start of one.

For waters less than 5 feet in depth, readings at the surface and at the bottom are usually sufficient. For waters more than 5 feet in depth, a reading should be taken at every 5 foot interval. For critical areas, however (at the top and bottom of the thermocline, for example), readings should be taken at 1 foot intervals.

The best method of obtaining a temperature series is with a *thermistor*, a small, battery-operated unit which usually has read-out scales in both °F and °C. A maximum–minimum thermometer can also be used, but this requires hauling the unit back to the surface each time to record the temperature and to reset the indicator markers. The last and least accurate method is to use a collecting bottle to obtain a water sample from the desired depth. Quickly raise it and take the temperature before it has a chance to change too much.

An indoor–outdoor thermometer of the type shown in Figure 5.5 is of real value in determining simultaneously the air temperature and the water temperature. Up to depths of 5 feet or so, this instrument can also be used to determine quickly and accurately water temperatures at different depths.

ig. 5.5
n indoor-outdoor ther-
mometer with a 5-foot sen-
or cord for use in ponds 5
et or less in depth.

5.13 VELOCITY OF FLOW

In Unit 4, you saw the extreme importance of velocity of flow. It influences many other factors such as D.O. concentration, carbon dioxide concentration, and temperature. Therefore a stream study would be incomplete without a determination of this physical factor.

Three methods are outlined here. Try each of these at least once before you settle on any one method for the duration of your studies. With Method 3 you can compare the velocities at different points within the stream. This is an important consideration to take into account when investigating stream life.

Method 1

Materials

a)   stop watch
b)   known length of string
c)   buoyant object such as an orange or a styrofoam ball

Procedure

Stand near the center of the stream and, having set the stop watch, drop the object on the end of the string into the water. At the moment the object hits the water, start the watch. As soon as the string becomes tight, stop the watch.

During this procedure, the hand holding the string should be as near as possible to the surface of the water.

You have now recorded the *time* and the *distance* (length of the string). If the string is measured in meters, you can very easily compute the velocity in *meters per second*. Repeat this procedure 3 or 4 times and average the results.

Method 2 THE THRUPP METHOD

Materials

a) straight flat bar of metal or wood

b) two nails or pegs

c) meter stick

Procedure

Set up the apparatus as shown in Figure 5.6. Place it in the water with the *nail end upstream*, holding the bar and meter stick parallel to and slightly *below* the surface of the stream. Ripple patterns will be set up as shown in Figure 5.7. As the stream velocity increases, the intersection of the wave patterns (B) will take place farther down the meter stick. Measure the distance AB in cm. Calculate the *surface velocity* in meters per second as follows:

$$\text{When } P_1P_2 = 10 \text{ cm, use } v = 0.1555 \, D$$
$$\text{When } P_1P_2 = 15 \text{ cm, use } v = 0.1466 \, D$$

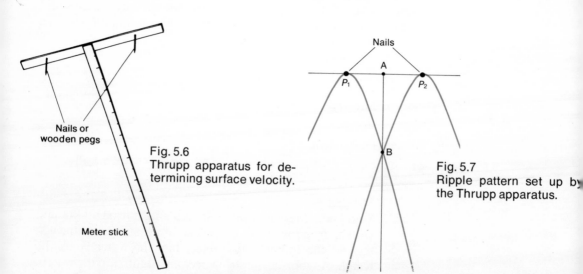

Fig. 5.6
Thrupp apparatus for determining surface velocity.

Fig. 5.7
Ripple pattern set up by the Thrupp apparatus.

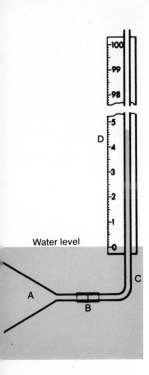

Water level

In these formulas, v = surface velocity in meters per second and D = distance AB in centimeters.

Note: This method will give you only surface velocity and cannot be used in very slow-moving streams.

Method 3

Materials

You can assemble your own apparatus as shown in Figure 5.8.

Procedure

Place the tube in the water as shown in Figure 5.8. The open end should point upstream so the water can enter. Take a reading from the scale and compare it to readings taken at other points in the stream. The actual calculation of the velocity is done with the formula

$$v = \sqrt{2gh}$$

where v is the *velocity* in meters per second, h is the *height* in meters to which the water rises in the tube above the normal level of the water, and g is the *acceleration due to gravity* (9.8 m per sec²).

This is one of the most useful methods for determining velocity at a point—in rapids, behind rocks, on the bottom, or at the surface. Compare animal life, plant life, water temperature, and so on at these various points.

5.14 VOLUME OF FLOW

Have you ever watched the seemingly endless flow of a river or stream? It is sometimes quite incredible how large a volume of water can flow past you, all going towards the same place. The method outlined here describes how to calculate the volume of flow of a river or stream. Also, by comparing the volume of flow into and out of a certain lake or pond, you can tell whether there is any ground seepage or whether an excessive amount of water is evaporating from the surface. These factors are of considerable interest to us during studies of lakes and ponds.

Materials

a)   stop watch

b)   float

c)   tape measure

Procedure

Determine the following values:

t—the time in seconds required for the float to travel a measured section of a stream

l—the length in meters of the stream section

w—the average width in meters of the stream section

d—the average depth in meters of the stream section

To compute the rate or volume of flow in cubic meters per second, use the formula

$$r = \frac{wdal}{t}$$

where a is a constant. The value of a is 0.8 if the stream bed is composed of rubble or gravel, and 0.9 if the stream bed is quite smooth (sand, mud, silt, or bedrock).

If you have already measured the stream velocity in meters per second by one of the methods outlined in Section 5.13, use the formula

$$r = wdav$$

where v is the velocity of the stream.

Problem: If you measure the volume of flow for the same stream in a very wide section and in a very narrow section, will the two values differ? Give your reasons for your prediction. Now measure the values. Was your prediction correct?

5.15 CROSS-SECTIONAL PROFILE OF A STREAM

Any stream study should include a determination of the cross-sectional profile of the stream. The profile may explain certain variations in temperature, velocity, and types of life across the width of the stream.

Materials

a) string

b) meter stick

Procedure

Suspend a string across the width of the stream, its ends tied securely at each side. At suitable intervals along the string, measure the depth of the water, the type of bottom material, and the amount of vegetation. (For a stream 1 m wide, a suitable interval is 5 cm; for a stream 5 m wide, 0.5 m.) In recording the bottom type, you can use the system below or you can construct one of your own.

Bedrock	————
Mud or Silt	··············
Rubble	********
Sand	xxxxxxxx
Gravel	oooooooo

When you have recorded your data, use a suitable scale to reconstruct the stream profile on paper.

5.16 DEPTH PROFILE OF A SMALL LAKE

It is often important to know the shape and the size of a lake basin. Not only does a profile show the irregularities of the basin, but it also provides information from which such things as volume can be calculated. It must be understood, however, that a depth profile changes often. Each year the water level fluctuates, thus changing the depth figures. In addition, constant filling in of the basin over a period of years (a successional change) also changes the profile.

The following experiment explains the methods involved in depth sounding a lake. The final product is a depth contour map.

Materials

a) boat

b) depth sounder or sounding lines

c) ropes

Procedure

a) No matter whether a sounder or sounding lines are used, the first step is to acquire a map of the lake.

b) On this map plot your probable sounding runs (lines connecting shore to shore) so that the best possible coverage of the lake is obtained (Fig. 5.9 A).

c) If a depth sounder is used, follow the instructions that accompany the instrument.

Fig. 5.9
These three diagrams illustrate how to construct a contour map of a pond or lake: (A) planning the sounding runs; (B) plotting depth readings on the field map; (C) joining the depth readings to complete the contour map.

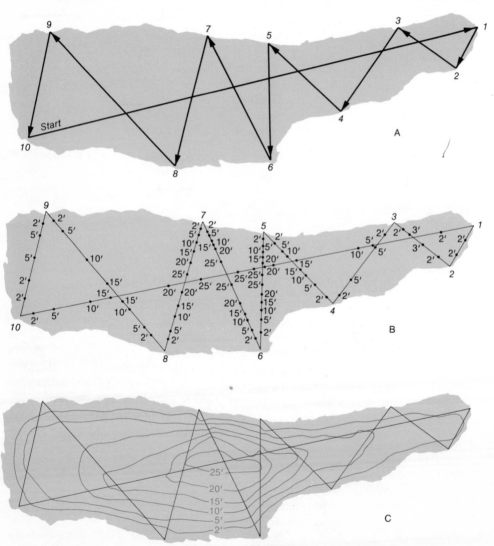

d)   If sounding lines are used, stretch a rope, marked in 10 foot intervals, from the start to the end of run 1. Pull the boat along, stopping at each mark to take a depth reading and record it on the map. Then stretch the rope along run 2, and so on (Fig. 5.9 B). After all the runs are complete, join equal depths with a line to get the contour map (Fig. 5.9 C).

5.17   A MODEL OF A LAKE

The temperature stratification (layering) of lake waters resulting from variation in water densities during the summer can be studied in the laboratory. The action of winds upon the thermocline can also be shown. Be sure to read Section 2.2 again as you perform this exercise.

Materials

a)   a clear glass oven dish or similar container

b)   a rubber siphoning tube

c)   an electric fan

d)   beaker

e)   a 2% salt water solution with red food coloring or water color

Procedure

a)   Fill the dish one-third full with warm tap water. Allow it to sit for a few minutes until currents stop.

b)   Set a container filled with cool colored salt water beside the dish. Elevate it slightly above the dish to permit siphoning.

c)   Slowly and gently, siphon the cool colored water into the bottom of the dish. Use a thin rubber tube.

d)   Allow the liquids to settle. Two distinct layers should be evident. The clear upper layer represents the epilimnion and the red bottom layer represents the hypolimnion. What does the interface between the layers represent?

e)   Set up a fan to blow "wind" across the surface of the "lake" for about 5 minutes. What happens to the thermocline? Will an offshore breeze create cool or warm swimming conditions?

f) Currents should develop in the epilimnion and hypolimnion (see Section 2.2). These can be shown by putting a few drops of ink into the water. Is the current flow on the bottom upwind or downwind?

g) If the thermocline breaks the surface, this "upwelling" will result in mixing. Therefore do not allow the fan to blow for too long.

h) Shut the fan off and look immediately for a short-lived surface oscillation. In lakes this is called a "surface seiche." Look for oscillations in the thermocline. These are called "internal seiches."

Discussion

Describe and explain the results of this study. How are similar effects produced in lakes? Try Case Study 7.

5.18 DETERMINATION OF WATER PRESSURES

One of the physical features of a lake that helps to determine the organisms present is the increase in pressure that accompanies an increase in depth. Also, the rate of descent or ascent of many organisms in deep lakes is partly determined by the changes in pressure at various depths. The phantom midge exhibits a daily migration from the bottom of deep lakes during the day to the surface at night. Changes in the size of air bladders in its body permit this migration.

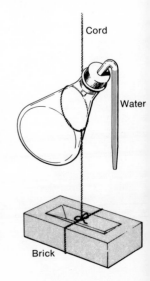

Fig. 5.10
Apparatus for determining
water pressure.

Cord

Water

Brick

The apparatus shown in Figure 5.10 should enable you to determine pressures at various depths. You may be able to devise some alternate forms of apparatus. The small constriction in the glass tube should help keep water in the tube as the apparatus is put together. The original volume (V_1) of the air in the container at atmospheric pressure (P_1) equals that in the flask and in the short section of tubing. When the apparatus is lowered into the water using a heavy weight or ballast, the increased pressure will decrease (compress) the volume of air in the container and water will enter through the tubing. When the container is lowered to the required depth a new volume (V_2) will be established equal to V_1 minus the volume of water now in the flask (W). Thus, $V_2 = V_1 - W$. This water will stay in the flask when it is raised to the surface. It can be measured in a graduated cylinder.

According to Boyle's Law, the product of the pressure (P) times the volume (V) of a given quantity of gas remains constant at a given temperature. That is,

$$P_1 V_1 = P_2 V_2$$

In this experiment we wish to determine P_2 at various depths and can do so using the following equation:

$$P_1 V_1 = P_2 (V_1 - W)$$

$$P_2 = \frac{P_1 V_1}{(V_1 - W)}$$

You may realize that the P_2 calculated will be partly in error since the temperature of the air will not be the same when submerged in the lake. The proper equation is actually:

$$\frac{P_1 V_1}{T_1} = \frac{P_2 V_2}{T_2}$$

A maximum-minimum thermometer will give you T_1 and T_2 (the temperature of the air at the surface and in the deepest water). The expanded equation becomes:

$$P_2 = \left(\frac{P_1 V_1}{T_1} \right) \left(\frac{T_2}{V_1 - W} \right)$$

Procedure

a) Set up the apparatus as shown in Figure 5.10. Determine V_1 by filling the flask and emptying the contents into a graduate.

b) Mark the cord off in 5 or 10 foot intervals.

c) Lower the apparatus 5 feet and then raise it. Measure the volume W and calculate $V_2 = V_1 - W$. Determine P_2 for 5 feet.

d) Similarly determine P_2 for 10 feet, 15 feet, 20 feet, etc.

5.19   A STUDY OF A MINI-ECOSYSTEM
(Also see Section 5.20.)

Frequently it is not feasible to go into the field to study an aquatic ecosystem. It is quite easy, however, to set up your own aquatic ecosystem in the classroom. There are several advantages to studying such a classroom ecosystem. First, the study area is small. Second, it is a relatively closed system; that is, it has few outside factors over which the experimenter has no control. Third, the experimenter can easily manipulate factors to determine their effects. On the other hand, there are several drawbacks to a system such as this. The most notable one is that it is a rather artificial system, and, in many ways, not at all like what is found in the field.

There are several ways of setting up a system for study. The easiest one is collecting a few gallons of pond water. The living organisms in this water should keep you quite occupied. For a more simplified study, however, a method of setting up an ecosystem involving *Daphnia* is described below. (See Sections 3.5 and 3.9 for information on *Daphnia*.)

Daphnia are useful in laboratory studies since they are easy to raise and large enough to be readily observed. They are also simple enough not to require a great deal of equipment for their maintenance. In most laboratory work, however, only one or two species of *Daphnia* are used, since too many parameters (for example, different ages of maturation, different sizes, and different tolerance ranges) would be introduced if more species were used. The calculations involved would become much too complicated.

In order to acquire a single species of *Daphnia*, various culturing techniques are used. For your purposes a simple technique is adequate. The best species to use is *Daphnia magna*, since these organisms are the largest and are quite active. The following experiment assumes that *D. magna* is used. (The techniques involved, however, can be used for most species.) In this experiment you will study the effects on *D. magna* of changing various chemical and physical factors.

Materials

a)   plankton net (170 or 180 meshes per inch) or a large jar

b)   sorting trays (white porcelain trays)

c)   eye droppers

d)   1 or 2 gallon aquaria (or 1 gallon jars) with glass tops

e)   light sources

f)   thermometers

g)   large glass jars (1–2 liters)

h)   spring water, filtered pond water, or tap water which has been treated with an anti-chlorine compound (available in pet shops) and allowed to stand for one week. The water should have a *p*H of about 7.

Procedure

A. To Start Cultures

Note: Use as many containers for your cultures as you wish, but always set up an extra one as a control (that is, no conditions are changed in it). This is the source of comparisons to be made with your other cultures.

a)   Obtain some *D. magna*. They can be bought from a biological supply house or collected with a net or bottle from a shallow pond, near the surface and away from currents.

b)   If you are collecting your own, take along a sorting tray, an eye dropper, a plankton net, and a large jar of the water which you are going to use in your cultures. Make sure that this water is at the same temperature as the water in the pond from which you are collecting.

c)   Collect zooplankton by pouring several jars of pond water through the plankton net.

d)   Empty the contents of the vial on the plankton net into the sorting tray. With the eye dropper (a hand lens might also be helpful), remove the largest *Daphnia* which have young in their brood pouches. Place them in the jar of water, together with some debris from the pond or its bottom. By taking only the largest *Daphnia*, you are almost ensuring that you have *D. magna*. Absolute certainty with this method is, unfortunately, not possible.

e)   When you have enough *Daphnia*, take them back to school. Let them stand in sunlight for about two or three days. This allows the temperature of the water in the jar to become the same as that of the water in your aquaria. It also allows the *Daphnia* to acclimate to the newer temperature. If you prefer, you can place heaters in all aquaria as well as in the jar, in order to keep the temperature at 22°C (72°F), the optimum temperature for *Daphnia magna*.

f)   Place several mature *Daphnia* into each aquarium. Allow them to grow in numbers for a week or so. Keep the aquaria well-lighted so that a slightly greenish color is always present (due to algae). This indicates good food conditions. All aquaria should be covered with glass.

g)   To maintain the cultures, add about ½-pint of pond water rich in algae and other microorganisms to each culture every week or so. This should be enough to maintain a continuous population which is balanced, since there is enough food to maintain only so many *D. magna*.

h) If a "skin" develops on the top of the water, skim it off or change the water.

i) If the glass covers are kept in place as much as possible, little water will be lost through evaporation. Thus there will be little need for adding more water to the aquaria.

B. Changing the Environment

When all cultures are doing well, each can be considered a separate ecosystem. From the viewpoint of population, each ecosystem is closed since the population size and food availability are in a constant state of balance. Each is a limiting factor: the *Daphnia* population limits the amount of microorganism growth; the amount of microorganism material present limits the population of *Daphnia*. To verify that the population has reached its maximum size, count the number of *Daphnia* in each container at 3-day intervals for 2 to 3 weeks. Although there may be fluctuations, the average population will remain the same. (This counting can best be done by stirring up the water and then drawing off, say, one pint of water and *Daphnia*. Count the *Daphnia* in this sample and multiply by 8, the number of pints in a gallon. This gives an approximate number only!)

What happens when we start to tamper with this balanced ecosystem? In other words, what effects can be seen when we change the factors discussed earlier in the text?

The results of all of the following tests should be recorded and compared to your control culture.

a) Lower or raise the temperature of a culture: slowly; quickly. Record observations on changes in:

    rate of heartbeat
    population levels
    D.O. levels
    swimming habits.

b) Reduce the D.O. level in a culture by cutting off the light source. Observe the effects on:

    color of the *Daphnia*
    swimming habits
    population levels
    carbon dioxide levels.

c) Increase or decrease the *pH* in a culture: slowly; quickly. Note the effects on:

    population levels
    heartbeat rate.

Increasing or decreasing the pH can be done by adding 0.1M base or acid solution, a few ml at a time, until the desired pH is reached. You may carry out this addition either slowly over one or two days or quickly over several hours.

d)   Increase the hardness of the water by adding a dilute solution of magnesium or calcium salts (or both): slowly; quickly. Observe the effects.

e)   Increase or decrease the food supply of a culture and note the effects.

f)   Aerate a culture and observe what happens.

For those who wish to do more, two other experiments are briefly described below. It is left to you to work out the details of the experiments.

The first involves the use of two competitive species of *Daphnia* in the same culture. The two normally compete for food and space. By altering the environment as in the above experiment, you can favor either one or the other species.

In the second experiment, cultures of one species of *Daphnia* are raised in the dark and fed every 2 or 4 days (with a fixed amount of algae culture). Every third day, a certain percentage of each culture is removed. Percentages which can be used are 0, 25, 50, 75, and 90. The effects of *predation* on population levels can be studied in this manner. You are the predator.

5.20   A STUDY OF A MINI-ECOSYSTEM (ADVANCED)

In this study you are, in essence, repeating the experiment outlined in Section 5.19, with one major modification. Also, make as many further modifications as occur to you in the course of the experiment. Our modification is this: Instead of restricting the animal life in your "indoor pond" to one species (as in 5.19), use all of the species that are present in a sample of natural pond water that you collect.

When you think of a pond, you undoubtedly imagine frogs, turtles, pond lilies, bulrushes, and so forth. By now you know that these macroscopic organisms constitute only a part of the life of the pond ecosystem. A myriad of microscopic organisms operate within the pond. Without them, the rest of the community could not exist. Among these important organisms are bacteria, blue-green algae, green algae, protozoa, rotifers, and small crustaceans.

In this study you are to collect a few gallons of representative pond water. Make sure that you get some surface water (including duckweed and the organisms that reside on and amongst it), some water from different depths, and, most important of all, some water from the space just above the bottom detritus. Include also 3 or 4 cups of the bottom detritus. Do not include macroscopic animals like fish or frogs. In such a confined volume, no balance could possibly be attained with such animals present. Do not reject, however, such things as nematodes, *Tubifex*, and planaria that may be in the bottom detritus. A few sprigs of some aquatic plants like *Chara* and *Elodea* can be included.

Back in the laboratory, divide this water and its contents as equally as you can among 4 or 5 containers. Each should have a capacity of at least 2 quarts. Select one of these as the control and duplicate in it as many of the features of the natural pond as you can. For example, it should not be aerated and the light quality and intensity should correspond to the average light conditions in the natural setting. (Figure out for yourself how to duplicate these conditions.) The life in this control should be examined every few days. Keep track of the species present and determine the relative sizes of the populations of the various organisms.

For each of the remaining containers of pond water, change *one* environmental factor. Monitor the effects that this change has on the life in the water. For example, you might try reducing the light intensity or aerating the water. Section 5.19 suggests further things that you can try. We are sure that you can think of still other experiments that you would like to try. Don't expect instant results. In some cases, no obvious changes will occur for several weeks or even for several months. This is a long-term experiment and should provide you with months of enjoyment at the microscope? Try to predict in advance the effects of the changes that you impose on the environment. For example, suppose that you aerate one container heavily. What effect do you think this will have on duckweed, a plant that appears to enjoy stagnant water? If the duckweed dies, what chain of events will be set off?

Remember that your main objective in this study is to determine relationships between abiotic and biotic factors in the pond ecosystem. You should also learn a great deal about food chains and food webs. You might even take time to study the morphology and physiology of some of the organisms that you find in the water.

5.21   PRIMARY PRODUCTION

The primary production of an aquatic ecosystem is the rate at which energy from the environment is utilized to form organic compounds through photosynthesis. To arrive at an index of this productivity, the amount of oxygen produced by part of the system is measured. Why is the oxygen measured?

This is an *index* only, and cannot be taken as a figure representing primary production. The actual process of photosynthesis varies with the rate of respiration of organisms, the light intensity, the photoperiod, the cloud cover, the seasons, the turbidity, the temperature, and other factors.

Materials

a)   500 ml bottles

b)   aluminum foil

c)   D.O. test kit

Procedure

a)   Determine the D.O. of a lake at various places and depths and record the values.

b)   Collect two bottles of water at each of these places and depths. If the water is rich in phytoplankton and zooplankton, the water is enough. Otherwise add some sprigs of larger aquatic vegetation to each bottle. Try to keep these sprigs fairly equal in size. Seal both bottles.

c)   Cover one bottle completely with aluminum foil.

d)   Suspend both bottles at the location and depth where they were taken (Fig. 5.11). In the uncovered bottle respiration and photosynthesis will continue, while in the covered bottle only respiration will occur.

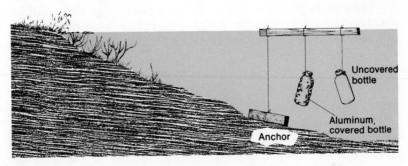

Fig. 5.11
The positioning of sample bottles for the determination of primary production.

e)   After a certain period of time (1 to 24 hours), each bottle is analyzed for D.O. and these values recorded. The total oxygen produced for the time interval is the sum of the differences between the D.O. content at the start and the finish in each bottle.

Example:

D.O. at end, uncovered	12.3 ppm
D.O. at start, uncovered	− 6.4 ppm
Net oxygen produced	5.9 ppm
D.O. at start, covered	6.4 ppm
D.O. at end, covered	− 1.9 ppm
Oxygen consumed	4.5 ppm
Net oxygen produced	5.9 ppm
Oxygen consumed	+ 4.5 ppm
Total oxygen produced	10.4 ppm

Thus an index of the productivity is the production of 10.4 ppm of oxygen in "x" units of time (or 10.4 mg of oxygen per liter of water in "x" hours).

5.22   THE EFFECT OF CROWDING ON A POPULATION

When you sample organisms in a pond or stream, you will most likely notice variations of size in different areas. This difference in size may be due to an increase or decrease in the population density. An extreme case of this is *crowding*. You are never likely to see a 4-inch goldfish in an aquarium which is 2 feet long. However, goldfish of this size are quite common in large outdoor ponds.

In this experiment you will study crowding in populations of bacteria. There are two reasons for using these tiny organisms. First, you can become acquainted with the bacteria which are always present in pond and stream water, and which you may not be able to observe otherwise. Second, because of their rapid reproduction rate, bacteria are very easy to work with in the classroom.

Materials

a)   50 ml flask of distilled water

b)   4 test tubes, each containing 9 ml of nutrient agar

c)   culture tube of bacteria (*Serratia marcescens*)

d)	sterile 1 ml pipette	h)	transfer loop
e)	4 sterile test tubes	i)	Bunsen burner
f)	2 sterile 10 ml pipettes	j)	hot water bath
g)	4 sterile petri dishes	k)	millimeter ruler

Procedure

a) After heating the water bath to 100°C, place the 4 tubes of agar into the bath.

b) With the 10 ml pipette, add 10 ml of distilled water to each of the 4 empty tubes. Label these tubes A, B, C, and D.

c) Add 1 loopful of the bacteria to the water in tube A. Mix this thoroughly by rolling the tube between your hands.

d) With the 1 ml pipette remove 1 ml of the mixture from tube A and add it to the water in tube B. Mix well and add 1 ml of this mixture to tube C. *Do not add anything to tube D.*

e) When the agar has melted in the tubes which are in the water bath, remove the tubes and allow the agar to cool to 45°C. Add 1 ml of the mixture in tube A to one of the agar tubes. Mix this quickly and pour it into one of the petri dishes. Move the dish in a "figure 8" pattern on the top of your desk to gently mix the contents. Label the petri dish "A." Now repeat this procedure with tubes B, C, and D.

f) Let the 4 dishes A, B, C, and D stand for a few minutes to solidify. Place the dishes bottom side up in a warm place for 24 hours.

g) After 24 hours, examine the petri dishes and determine the number of bacterial colonies on each. Record your data in a table similar to the one below.

Petri dish	Number of colonies		Average diameter of colonies		Average area of colonies		Total area of bacterial growth	
	Day 2	Day 3	Day 2	Day 3	Day 2	Day 3	Day 2	Day 3
A								
B								
C								
D								

h) With the millimeter rule, determine the average diameter of the colonies to the nearest 0.5 mm. Then compute the average area of the colonies. Record these data in your table.

i) Place the dishes in the dark for another 24 hours.

j) Repeat steps (g) and (h).

Discussion

Describe and account for your results. Is there any evidence of the effects of crowding in the mini-ecosystems that you set up (Sections 5.19 and 5.20)? What effects might population density have on the size of the fish in a small lake? Why?

5.23 LABORATORY INVESTIGATION OF BREATHING MECHANISMS

Oxygen is the one gas that animals are vitally dependent upon for their life processes. Extremely small organisms living in the water usually do not possess any special organs for obtaining oxygen. They rely upon the diffusion of the gas directly into their body tissues from the water. Larger animals, especially those with hard outer "skins," must have special adaptations to bathe their body tissues in the needed oxygen. Blood and circulatory systems, gills and lungs, are all mechanisms or adaptations to aid in this process.

Aquatic insects are an interesting group for the study of breathing mechanisms. Insects first evolved from the marine environment to live on land. As a result, they developed systems for breathing air directly. Only a few insects have since invaded the aquatic world. In some cases, these species have developed new means of breathing. Of course, some forms still breathe air directly at the surface. Others are independent of the surface, and obtain dissolved oxygen from the water. Still other forms obtain oxygen both from the surface and from the water, and are not exclusively dependent on either.

Dissolved oxygen (D.O.) can vary widely in different bodies of water, at different times of day, and at different times of the year. D.O. is thus an important limiting factor for those organisms that must obtain their oxygen from the water. This experiment shows not only where different aquatic organisms obtain their oxygen, but also which organisms can survive in environments having low concentrations of D.O.

Materials

a) 1 oxygen gas cylinder

b) 2 large glass containers

c) glass and/or rubber tubing

d) 1 nitrogen gas cylinder

e) 4 2-hole rubber stoppers

f) 4 glass flasks

g) minimum of 4 water striders; 4 water boatmen; 4 guppies; 4 dragonfly (or damselfly) nymphs

Procedure

a) Boil pond water or dechlorinated tap water to remove the dissolved gases present. Allow the water to return to room temperature. Pour equal amounts of this water into two different containers. (A small amount might be tested to determine the number of parts of oxygen per million parts of water present.)

b) Bubble nitrogen gas from a cylinder through the water in one container. Bubble oxygen gas from a cylinder through the other. Bubbling many small bubbles will saturate the solution faster than an equal volume of large bubbles. This will take 10 minutes or so.

c) Prepare two set-ups like the one shown below. Pass a very slight flow of oxygen gas through one apparatus. Pass a very slight flow of nitrogen gas through the other.

d) Each flask should receive one or preferably two specimens of each organism to be tested. Suitable organisms are the

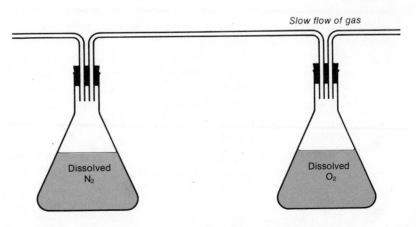

Fig. 5.12
Apparatus for a study of breathing mechanisms of aquatic insects.

water strider, water boatman, dragonfly or damselfly nymph, and guppies. Organisms that could be tested in other experiments include the predaceous diving beetle larva and adult, the water scorpion, and any of the crustaceans. Even zooplankton could be tested to determine their response to the various conditions.

e) Observe the organisms for 30 minutes or until 100% of them are incapacitated. The inability to swim, to remain erect, or to show any movement are criteria that you can use to judge when the organisms are incapacitated. What other criteria can you use?

f) For each species record the time until 50% and 100% of the specimens are incapacitated.

g) Results can be compared with others in the class. Calculate average times.

h) Leave the apparatus set up for 24 hours, if space is available, to observe organisms unaffected after 30 minutes.

Discussion

After categorizing the test organisms as to their mode of respiration, discuss the habitat conditions that you think these animals are best adapted to. If oxygen is not a limiting factor (it is always in adequate supply) for air breathers, why haven't all aquatic forms maintained a way of breathing at the surface?

5.24 PLANKTON FILTRATION IN THE FIELD OR LABORATORY

Much of the drinking water supplied to the large cities of North America (and of the whole world) must come from large lakes and rivers where a continuous supply is available. The water from these sources contains plankton, some of which may impart disagreeable tastes and odors to the water. In addition, suspended silt is present which must be removed. The Sedgwick–Rafter method for filtering small water samples employs the same principle used in the water filtration plants of large cities. Water is passed through fine sand and the suspended material is removed.

Procedure

a) Set up the apparatus as shown in Figure 5.13. A special Sedgwick–Rafter tube has been devised for this filtration

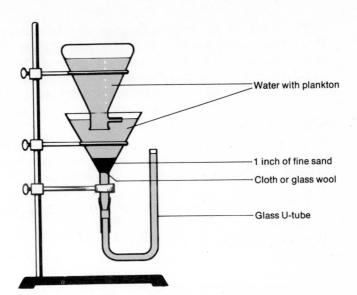

Water with plankton

1 inch of fine sand

Cloth or glass wool

Glass U-tube

method, but a wide variety of modifications can be substituted. An ordinary glass funnel about 8 inches in diameter can be used. The sand filter bed can be supported by the glass wool used in aquarium filters or by a piece of cloth. The sand should be fairly fine and should be sorted through a fine mesh net or screen.

b) A couple of inches of distilled water should be placed in the tube or funnel before adding the sand. If the sand is slowly introduced to a thickness of about 1 inch, it will not have pockets of air trapped in it. More distilled water may be used to wash the sand into place. This water can be drained off.

c) Gravity filtration of this type may take 30 minutes to filter a 500 ml sample unless some form of suction is used. A long siphon can be added to the end of the funnel. The gravity pull on this added column of water will speed up the process. Suction should not be too great, otherwise the plankton might be injured.

d) To retain a small amount of water in the funnel, loop the siphoning tube over the support apparatus. This stops the siphoning process. If a siphon is not used, a U tube will prevent all of the water from draining through and injuring the plankton.

e) The sand and remaining water should be transferred to a small beaker. The funnel can be rinsed with 5–10 ml of water from a pipette. By gently mixing the sand and water in the beaker, the plankton will be resuspended. The water should be quickly poured into a second beaker. By adding another 5 ml to the sand and again mixing, most of the remaining plankton can also be removed. The normal practice in a filtration plant is to backwash the sand and dispose of the backwash water and its silt and plankton.

f) Obtain a drop of the backwash water in an eye dropper and place it on a glass slide. Place a drop of the filtrate on another glass slide. Count or estimate the number of organisms seen in one minute at an arbitrarily selected spot in each drop of water. Observe at about 100x magnification.

Discussion

Account for the differences observed between the backwash water and the filtrate. Is the filtrate suitable for drinking?

5.25   LABORATORY STUDY ON ANIMAL BEHAVIOR

To understand how an organism survives in its environment, you must know something about its behavior. The tiny crustacean *Daphnia* can be found in ponds and lakes in great numbers. It can be kept easily in the laboratory by setting up an infusorium that cultures protozoans for food (see Section 5.19). The water flea shows some interesting behavior in response to environmental conditions. This experiment illustrates the behavior that occurs in response to two factors, carbon dioxide concentration and light. This behavior probably has some advantage to *Daphnia*, increasing its chances of survival. Your problem is to discover the behavior and then to hypothesize the advantage gained with it.

Procedure

a) In a room illuminated by dim uniform light, set up two glass bottles containing well-aerated, non-chlorinated water.

b) Into one container bubble carbon dioxide from a gas cylinder for 5–10 minutes.

c) Transfer approximately equal numbers of *Daphnia* from a stock solution into the two containers.

d) Observe the distribution of the *Daphnia* in the containers for a period of 3–5 minutes.

e) Provide a light source that illuminates the bottles from above. Observe the containers for 3–5 minutes.

f) Change the position of the light source and illuminate the bottles from the sides and the bottom, if possible. Observe the distribution for each change in position.

Discussion

Some ecologists believe that *Daphnia* learn that, when carbon dioxide concentrations are high in the water, the greatest amounts of available oxygen will be found close to the surface. It is true that at the surface oxygen diffuses from the air into the water. Actually, *Daphnia* seem to learn that they must go in the direction of the light source to find the surface (where the highest oxygen concentration is). This would work in nature, but in experiments the light can be adjusted to come from any direction. Do your results correspond with these ideas? When CO_2 concentrations are high do *Daphnia* always travel towards the light source? Is the attraction as great in well-oxygenated water?

5.26 CLASSROOM EXPERIMENT: LIEBIG'S LAW OF THE MINIMUM

Phosphorus (P) and nitrogen (N) are necessary nutrients for the growth of any plant life. The production of carbohydrates, through the process of photosynthesis, does not in itself satisfy the needs of plant cells. Various fats and proteins must be formed from carbohydrates to furnish the additional compounds necessary for life. Phosphorus and nitrogen enable this to take place. But, if either of these nutrients is in short supply, it becomes a limiting factor, retarding further growth. In chemistry, it is stated that when the reagent in minimum supply has been used up, the reaction must end. This is Liebig's Law of the Minimum.

Materials

a) eye dropper
b) water bottles or some other glass containers
c) teaspoon
d) tablespoon
e) suspension of green aquarium scum
f) liquid plant fertilizer

g)   distilled water

h)   half-inch pieces of chalk

Procedure

a)   Set up 7 bottles each containing a piece of chalk and 10 drops of scum suspension.

b)   Fill the first bottle with liquid fertilizer, and to the others add 5 tablespoonfuls, 1 tablespoonful, 1 teaspoonful, 20 drops, 2 drops and to the last add none. Then, fill the last 6 bottles with distilled water.

c)   Place a cardboard dust cover over each bottle and place them near a light source.

d)   This is a long term experiment that should be observed at intervals for a period of weeks or months.

e)   To make periodic checks of the growth of the scum, tie a glass slide to a piece of string. The slide can be dangled in the water by tying the other end of the string to the cardboard cover. The slides should be removed every week or so for microscopic examination.

f)   You might try turning the bottles 180 degrees in the light source to observe the effects of shading by the growth of scum on the side closest to the light source.

Discussion

How does the quantity of liquid fertilizer relate to Liebig's Law of the Minimum? Why might the growth of scum on the slides be different in quantity from that on the sides of the container? What are the effects of shading? Why?

5.27   LABORATORY STUDY: ANALYZING GUT CONTENTS

The things that animals eat tell a great deal about their roles in the communities in which they are found. For years, all sorts of animals, from elephants to fleas, have been analyzed for their stomach contents by ecologists interested in food chains. The large carnivorous, omnivorous, and herbivorous insects, crustaceans, and fish can be collected for the express purpose of analyzing their gut contents.

Care must be taken in removing the alimentary tract of an insect. A dissecting microscope is of great use. For many of the insects, merely cutting off the head and the end of the abdomen frees the gut. Remove it with forceps. Check the anatomy of a crayfish and of a real fish in a book before attempting a dissection of either of these animals.

Contents of the posterior section of an alimentary tract will be digested and unrecognizable. The recognizability of certain food items at the anterior end will depend on the manner in which the food was consumed. Some insects practically swallow their prey whole, as fish do. Others chew it to a fine mulch. Fine material should be placed on a slide and observed under the compound microscope. Fresh killed, "unpickled" specimens are best. Look for diatoms and other plant material, as well as for animal remains. Try to determine the diet of a particular species by analyzing a number of gut contents. Hypothesize its position in food chains.

5.28 CLASSROOM EXPERIMENT: HABITAT SELECTION

Animals exhibit behavior that results in their own selection of specific environmental conditions within which to live. Some animals show very little selectivity with respect to their "home." Others seem to choose on the basis of a single factor, while still others choose on the basis of many factors. Experiments can be set up in the laboratory, especially for pond-dwelling animals, to determine if they are indeed selective with respect to their environment.

Window glass
(cover)

Materials
a)   clear glass oven dishes
b)   flat black paint
c)   sand
d)   cardboard
e)   window glass
f)   loam
g)   gravel
h)   collection of pond organisms (compatible species including some burrowing forms)

Procedure

A. Habitat Color, Light vs Darkness
a)   Paint one half of an oven dish and window glass on the outside. Paint lengthwise, blackening only one side of each (Fig. 5.14).

Glass oven dish

g. 5.14
abitat selection.

b) Half fill the dish with pond water and place pond organisms in the dish. Place the window glass on the oven dish, with the blackened side of the glass over the blackened side of the dish.

c) Allow the container to remain overnight with a light source shining from one end.

d) Observe and list organisms as light-seeking, indifferent, or darkness-seeking.

B. Habitat Color with Cover

a) Set up as in the previous experiment. Add clear cellophane, crumpled and immersed in one end of the dish. The cellophane should extend the width of the dish so that it is half in the dark and half in the light. Clean marbles or stones will hold the cellophane down. Replace the glass cover.

b) Leave overnight as in previous experiment.

c) Some organisms may respond to the "need" for something to touch, although showing no response to light. They would be listed as cover-seeking but indifferent to light. There are five other possible categories which you should be able to list.

C. Habitat Color with Substrate Choice

a) Cut two pieces of cardboard to fit widthwise across the oven dish. These strips should be about 2 inches in height when taped into place at the edge of the dish. The dish should be now divided into 3 separate compartments of equal size.

b) Place gravel to a depth of 1 inch in an end compartment.

c) Place sand to a depth of 1 inch in the middle compartment.

d) Place some loam in a bucket with water and stir the mixture. Pour off floating substances after the mixture has had time to settle. Also pour off the rest of the water. Skim off the layer of fine silt and other material at the top of the settled soil. Place this material in the final compartment, adding enough water to evenly distribute it to a depth of 1 inch.

e) Remove the cardboard partitions carefully. Slowly add water to the container from the gravel end to raise the water to a depth of about 1 inch above the substrate.

f) Add the pond organisms.

g)   Place the glass cover lengthwise, darkening half of each substrate.

h)   Once again give the animals a day to orient themselves before categorizing them according to their substrate selection.

Discussion

Do the responses of the various animals correspond with the types of habitat in which they were originally found? Why is it that, in nature, sandy substrates invariably have a rather sparse aquatic invertebrate population?

5.29   AN EXPERIMENTAL FOOD CHAIN FOR THE LABORATORY

It is sometimes difficult to visualize a food "chain" in action. We exist in a three-dimensional world and organisms feed in a three-dimensional environment. When we think of a "chain" we are thinking of something one dimensional. You can closely approximate a natural food chain with a model such as the one shown in Figure 5.15.

The experimental food chain consists of three nearly independent environments. In nature they would be combined into one. This is an artificial representation of a natural process. Even so, it may give you another insight into the complex field of ecology.

Materials

a)   1 aerator-pump

b)   4–5 large containers (the second infusorium is not necessary)

c)   1 funnel

d)   glass tubing

e)   1 small fish (a baby guppy, capable of growth, is best)

f)   plankton sample

g)   grain

h)   egg yolk

i)   hay

Procedure

a) The apparatus is set up as in Figure 5.15. Note that the rate of transfer of organisms from tank to tank depends directly on the air bubbles carrying water over from the first reservoir. Too fast a rate of transfer may deplete the populations in the infusorian tank or in the crustacean tank.

b) The infusorian population can be started before the collection of planktonic crustaceans. A mixture of hay seed, wheat, rice, or some other grain, plus some short pieces of hay or grass should be boiled for 5 minutes. Part of the yolk of a hard-boiled egg could be crumbled and added when the mixture returns to room temperature.

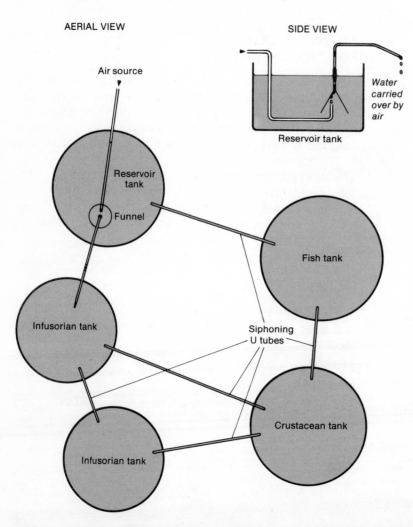

AERIAL VIEW

SIDE VIEW

Air source

Water carried over by air

Reservoir tank

Reservoir tank

Funnel

Fish tank

Infusorian tank

Siphoning U tubes

Infusorian tank

Crustacean tank

Fig. 5.15
A laboratory study of food chains.

c)   Visit a pond or lake and collect the following: some bottom sediment, a gallon of water that has been poured through a plankton net, some planktonic crustaceans like *Daphnia* or *Cyclops*, and some aquatic vegetation like *Elodea*.

d)   After returning from the field trip, add a small amount of the bottom sediment to the infusorian tanks. You might also add some of the water poured through a plankton net.

e)   The fish should be very small, a guppy or a small stickleback. Both are fairly tolerant species.

f)   A bit of aquatic vegetation like *Elodea* could be placed in the crustacean and fish tanks.

g)   Many variations are possible to this basic set-up.

h)   Observe the apparatus periodically over a period of weeks or months. Check to see that no link in the food chain is depleted severely. Record any changes that occur.

i)   Check the fish for growth.

Discussion

Scientists build mental or physical models in an attempt to simplify the rather complex happenings in real life. The obvious shortcomings of a model may bring into focus a rather hazy problem or concept. For instance, do you understand what is meant by the "balance of nature"? If so, does this food chain represent a "balanced" ecosystem? What do you think would happen if the model were changed? What series of events would take place if, for instance, the fish died? Suppose the fish eventually had young and these young could swim through the tubing from tank to tank. What series of events would take place then? With your knowledge of aquatic life, can you develop a more complex model that might function for years undisturbed?

5.30   SPAWNING BEHAVIOR OF FISH

Spawning behavior, like other behaviors, is the product of many biochemical changes. Ultimately, these changes must be initiated by the environment. A critical combination of water temperature and photoperiod or period of darkness will trigger spawning behavior in most species of fish.

       Since temperature and light are so important, it is obvious that spawning must take place during certain seasons. These are usually the spring or the fall. Once these conditions are met,

TABLE 2 SPAWNING CHARACTERISTICS OF SOME SPECIES OF FISH

Species	Season	Migration	Nest Description	Depth	Temp.	Parental Care
Sturgeon	spring	from lakes to streams	mud bottoms	shallow	12–15°C	none
Alewife	late spring	from deep to shallow water	sand bottoms	shallow	cool	none
Brown Trout	fall	not large or far	depressions in gravel bottoms of headwaters of streams	shallow	cool	none
Rainbow Trout	early spring	upstream to clear, fast water	gravel cleaned by fish	shallow	cold	none
Brook Trout	fall	upstream to headwaters	shallow depression; eggs covered by gravel	shallow	cold	none
Lake Trout	fall	from deep lakes to shallower waters	on rocky shoals or reefs	deep	cold	none
Lake Whitefish	late fall	usually none	rocky, gravelly, or sandy shoals	shallow	cold	none
Smelt	early spring	upstream	sandy and rocky areas of streams	shallow	cold	none
Northern Pike	early spring	usually not far	weedy, shallow bays	shallow	cold	none
White Sucker	spring	from lakes to streams	gravelly shoals	shallow	cold	none
Smallmouth Bass	late spring	none	shallow, clean depression in rocky bottom	shallow	15–20°C	male guard until fry are 1–2 weeks old
Largemouth Bass	late spring	none	mud or mixture of mud and organic matter	shallow	15–20°C	as for smallmouth bass
Yellow Perch	spring	to shallow waters	eggs in rope-like strands of mucus in no definite nest	shallow	5–10°C	none

spawning can occur only if a favorable site is present. Each species usually has an optimum site that meets the necessary conditions of light, temperature, food, water depth, and substrate.

Although no attempt is made to describe the spawning behavior for all species of fish, the accompanying table outlines it for some of the more common species.

Excellent observations of spawning fish can be made during fall or spring. Note their behavior and their color.

If you are lucky, you may find a nest. Note the number of eggs and see if you can figure out why such numbers are laid. What would happen if each egg survived? How are populations kept down to a proper size?

Note the parental care given to the eggs; to the young. Is there any? Compare presence or absence of this care to the number of eggs in the nest. What is the relationship?

Trace the position of the eggs and fry (young fish) in the food web of the ecosystem. What species depend on them? What would happen if fewer than the normal number of eggs were produced? Discuss this last question with respect to pollution and its effects on the reproduction of fish.

Most fish spawn at a particular time of year. How would you go about reversing the spawning cycle of a particular fish species so that, for example, a normal spring spawner becomes a fall spawner? What possible applications would this have?

Table 2 lists the spawning characteristics of some species of fish.

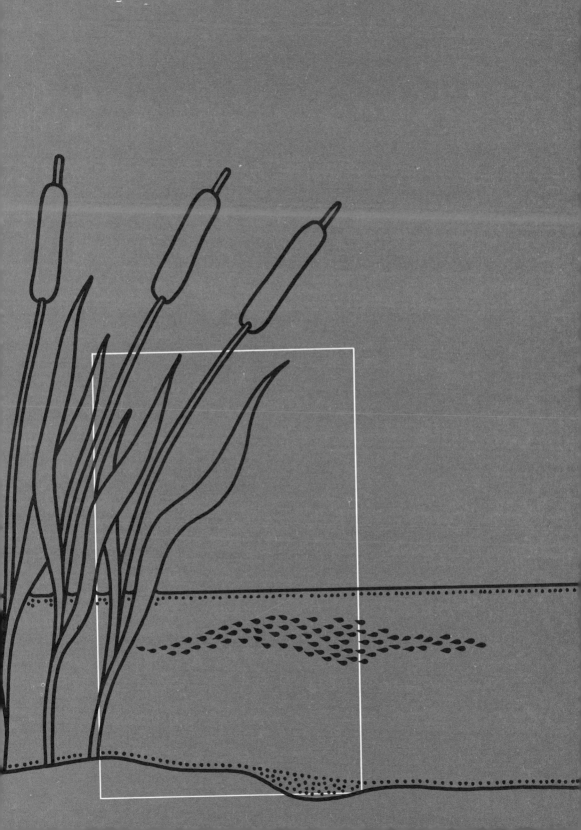

Major Field Studies

6

A. EQUIPMENT AND TECHNIQUES FOR BIOLOGICAL SAMPLING

This section deals with basic equipment and techniques for the successful field collection of biological material. Although in many cases the equipment is simple, the fact is, it works! With a little care and patience, the most inexpensive device can yield great dividends. In some cases, the equipment described is not a marketed product; your own handiwork will be needed to put it together. You may think of a better method of sampling the organisms. If so, you will need to design your own equipment. Scientists have to do this, and often advances in science go hand in hand with advances in technology.

6.1 THE SORTING TRAY

When an aquatic organism is disturbed, it usually has some means of movement and escape. But, movement also gives away the position of an animal if its body color does not match the background. Normally, aquatic organisms are well camouflaged, even in a sample of vegetation or bottom material removed from the water.

. 6.1
uatic organisms show
well against the white
ckground of a sorting
y.

A white sorting tray should be used in the field to locate, isolate, and capture individual specimens (Fig. 6.1). Either a white plastic or porcelain tray will do. It should be about 5–10 cm deep (to prevent spillage) and have about 1,000 square cm of surface area on which to sort. When transferred to the tray, the sample should occupy less than half of the total area. This still gives you over half of the tray in which to sort. Only about ½ of a cm of water need be in the bottom. The tray must be placed in a stationary position. Then any movement observed can only originate from a living organism.

6.2 OTHER SORTING EQUIPMENT

One pair of **forceps** should be attached with a string to each sorting tray. In this way they should not get lost. An **eye dropper** can be used to suck up minute, delicate worms or crustaceans. By heating and stretching a piece of glass tubing to a fine point, a very precise **micro-dropper** can be made for transferring tiny, individual organisms for microscopic examination. The micro-dropper may not require a rubber bulb to suck the organisms into the tube. Merely hold your finger over the open end until the tip is in position to capture the organism. Then remove your finger. The capillary action of the water moving up the tube will suck the organism in.

6.3   COLLECTING NETS

(a) A **kitchen sieve** is an ideal collecting tool (Fig. 6.2). It is most useful in "weedy" areas where something sturdy is needed to remove animals from the vegetation. The sieve is drawn back and forth through the water and aquatic plants until a sample is ready for sorting. The animals caught in the sieve will not usually move once out of the water. The sieve may be held partly in the water to observe movement, or the contents can be placed in a sorting tray. To transfer the sample to a sorting tray or other container, give the edge of the sieve a sharp rap on the side of the container. Samples of mud can be sifted at the surface of the water. The fine silt and debris can be washed from a mud sample, leaving the organisms and larger detritus. Agitate the sample at the surface, vibrating the sieve up and down so that fine particles wash through the holes.

Fig. 6.2
A kitchen sieve can b
used to capture sma
aquatic animals.

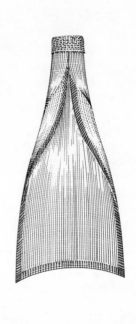

0.5cm

20cm

10cm

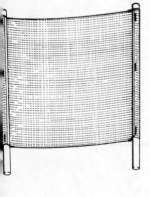

(b) You can make a useful **lifter** for collecting organisms and for transferring them from container to container (Fig. 6.3). The lifter is not a heavy duty apparatus, but is cheap and extremely practical for catching individual organisms seen scurrying through the water. It can be used in place of a sieve.

(c) The **hand screen** (Fig. 6.4) is used mainly for stream studies. It consists of two handles and a wire or fiberglass screen. If the net is made of fiberglass window screen, copper wires should be threaded through the top and bottom for extra strength. The ends of the wire or fiberglass should be stapled firmly to the handles with heavy duty staples or, better, placed into slots cut the length of the handles which are then nailed closed.

If a large 1 x 2 foot hand screen is built, two people should work jointly in making collections. One person holds the hand screen in the current, in a shallow section where the screen reaches from the bottom to the surface. The other person goes directly upstream a few feet to turn over stones and to wipe organisms from the rocks by hand. The current will carry the dislodged plants and animals down to the screen. The force of the water should trap them on the screen until it is removed from the water. Some organisms can be transferred to a sorting tray by holding the screen face down over the tray. Others can be washed from the screen by pouring water through it into the tray.

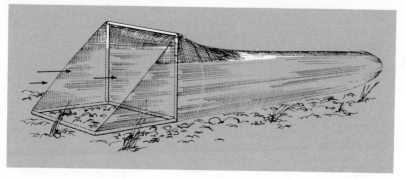

Fig. 6.5
A Surber sampler. Org[anisms] on stones and o[ther] bottom material within [a] square-foot frame are [dis]lodged by hand. The c[ur]rent then carries the [or]ganisms into the net.

A narrower 1 x 1 foot screen can be operated by one person. Hold both handles in an inverted V shape while turning rocks with the other hand. The kitchen sieve can also be used for this type of collecting. **The Surber square-foot stream bottom sampler** (Fig. 6.5) is a sophisticated apparatus used for quantitative sampling.

(d) For capturing tiny free swimming and drifting organisms, a fine-meshed **plankton net** is used (Fig. 6.6). The net is conical in shape, often with a small vial or container at the end of the cone. Normally the net is dragged by a rope behind a boat in the open water of a lake. The net can also be rigged to a pole and drawn back and forth by someone standing on the shore or in shallow water. In a stream, the net can be held in one place with the current flowing through it. In any case, avoid stirring up the bottom silt, because the net will get clogged with it.

You can construct your own plankton net out of any finemeshed cloth such as a woman's slip or a nylon stocking. To remove organisms trapped on a homemade net like the one in Figure 6.7, first turn the net inside out. Place a small amount of water in a flat container. Touch the net to the water. Many of the organisms will swim off the net on their own, while others can be gently washed free. Remember that the less water you use in this transfer, the more concentrated will be your plankton sample.

In weed beds, close to shore, large plankton populations occur. The **Birge cone net** has been devised to sample such areas. A wire screen prevents large plant material from clogging up the net while still allowing the plankton to enter. Exactly the same results can be obtained using a bucket, a kitchen sieve, and the regular plankton net all at once. One person scoops buckets of water from the "weedy" area and pours it through the sieve. The second person holds the sieve and the plankton net, the former above the latter. Large bits of vegetation and other material are trapped in the sieve; the plankton is carried through into the net. Many bucketfuls

Fig. 6.6
A plankton net provide[s a] concentrated sample [of] planktonic organisms.

may be required to get a concentrated plankton sample. You will recognize a fairly concentrated sample when the water begins to take a long time to drain from the net.

(e) The **dip net** is similar to the kitchen sieve in its uses. Because it has a long handle, it is most useful in deeper water. Sweeping this net through the vegetation will fill it with aquatic plants as well as animal life. Sampling of aquatic plants is thus possible, though the sorting of animal life will still be a manual exercise in a sorting tray. Since the cloth netting is light and fairly coarse-meshed, it is possible to use the dip net to catch fish. It might come in handy in the confined conditions of a stream.

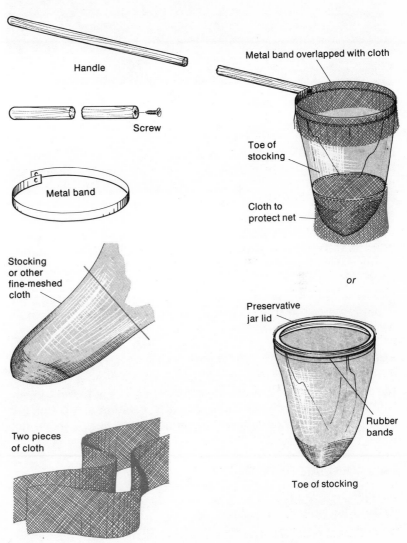

Fig. 6.7
Making your own plankton net.

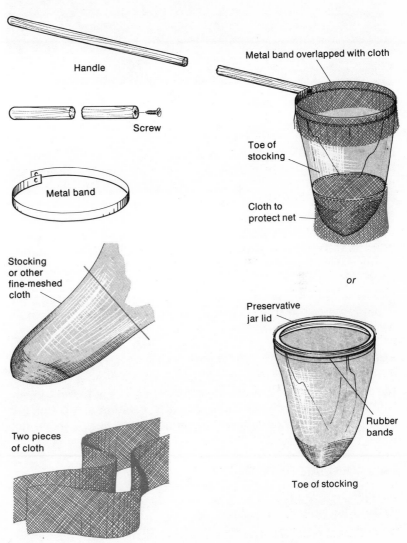

Fig. 6.8
The straight edge of the
frame dip net permits sa~
pling close to the bottom

Dip nets can be purchased in different shapes and sizes. The D-frame dip net shown in Figure 6.8 is very useful for sampling close to the bottom of a pond or stream.

(f) The **apron net** is specially constructed for sampling among the submergent aquatic plants around the shallow shore areas of a lake. The collecting container is rather elaborate, with a strong metal frame and various types of screening enclosing it (Fig. 6.9). It is attached to a long wooden pole. Because the apron net is pointed at the far end, it can be forced through aquatic vegetation quite easily. The top surface of the container has a coarse-meshed screen which allows small aquatic animals, dislodged from the plants, to fall into the container. At the same time it prevents the aquatic vegetation from entering. The bottom surface has a fine-meshed screen that traps the animals until they are removed.

Just as the net is taken from the water, you must give a last thrust with the pole. This should carry all the trapped organisms to the section of the net nearest the handle. A small door on the top surface is located at this end of the container so that the specimens can be removed easily. Some trash and vegetation will inevitably enter; often it is best to dump the contents into a white sorting tray.

Fig. 6.9
The sturdy construction
the apron net makes sa~
pling in weedy areas po~
sible.

6.4 EKMAN DREDGE

This piece of equipment is used to sample the organisms living in the bottom mud of lakes. Since the deep central sections of most lakes are covered with silt and sediments, often 50% or more of a

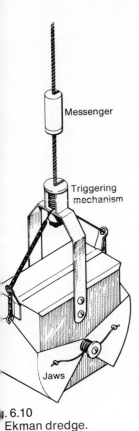

. 6.10
Ekman dredge.

lake's bottom is inhabited by mud-dwelling organisms. With the Ekman dredge and a boat, a couple of workers can randomly or systematically obtain samples of the bottom mud and inhabitants, from any depth.

The dredge is essentially a square brass box, with flap-like doors at one end and two large jaws at the other (Fig. 6.10). The jaws are kept wide open by chains as the dredge is lowered by rope. The upper doors open themselves as the dredge travels to the bottom, thus offering little resistance to the water passing through the box. When the dredge reaches the bottom, the operator should raise it a foot or so, hold the rope for about 10 seconds until it straightens out, then release it completely. The dredge should fall straight into the bottom material, nearly burying itself in the sediments.

The point of attachment of the rope is also the point of attachment of wire chains. These keep the jaws pulled open against the force of powerful springs that would otherwise clamp them shut. The chains are trigger-set to be released when a "messenger," a heavy metal weight, is dropped from the surface. It travels down the rope and strikes the triggering mechanism. When the jaws are released, the dredge is hauled up to the surface. On the way up, the top doors stay shut so that the sample remains undisturbed. Care must be taken while lifting the dredge from the water to avoid spillage. A bucket should be held in the water and the dredge placed directly into it. By adding water to the bucket, stirring the mixture, and pouring it through a sieve, the silt can be removed and the organisms collected. If a screen sifter is available, it can be used as an alternative, to sift and sort mud samples.

6.5 VEGETATION SAMPLING

Much of the vegetation can be sampled by hand although in deeper water other equipment might be needed. A simple garden rake can be used in most instances. A weighted ring of barbed wire attached to a long rope can be thrown out into "weed" beds when a boat is not available.

6.6 THE BAG SIEVE

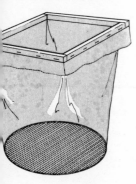

. 6.11
omemade bag sieve.

This piece of homemade apparatus is really an extra. It is useful for quick processing of samples of aquatic vegetation to remove animal life (Fig. 6.11). It is just a bag-like container that will float in the water. Various samples of vegetation are washed inside to remove clinging organisms. The screen at the bottom is made of fiberglass window screening, circular in shape and sewn around a bent coat hanger wire. A square wooden frame is sewn into place at the top to keep the bag floating when used in deeper water. The wire and wood keep the bag in shape, allowing the operator to have both arms free to collect and wash vegetation.

6.7   THE SIFTING SCREEN

This easily constructed piece of equipment is helpful for quick sorting of bottom mud samples (Fig. 6.12). The sample of mud is placed in the screen. The wood sides should keep the screen floating at the surface. Holding the sides in both hands, vibrate the screen and mud at the surface. The fine silt and mud particles quickly wash out the bottom while the organisms remain on the wire mesh. Large amounts of mud can be processed in this way. A kitchen sieve can be used equally well for small samples.

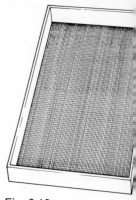

Fig. 6.12
A sifting screen that
about 35 cm x 25 cm
easily handled in
water.

6.8   THE MINNOW SEINE NET

These large nets are used primarily for capturing minnows and other small fish. If larger fish are foolish enough to be around at the time, they too can be caught since the nets are very strong. Minnow seines are about 4 feet deep and 6, 10, 20 feet or more in length. Two poles, attached at either end, are used to pull the net through the water. Lead weights or a chain keep the bottom edge down while floats keep the other edge at the surface. A one-eighth or one-quarter inch square mesh net will catch most fish. Any small enough to get away would be too small for identification anyway.

The proper method of holding the poles for normal stream use is shown in Figure 6.13. A number of problems are often encountered. The fish will escape underneath a net not held close to

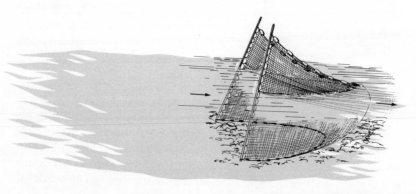

Fig. 6.13
Proper position for a se
net.

the bottom. Or, if the net gets caught in vegetation or on a twig or rock, the fish can again escape. It may not take long before you begin to wonder about the I.Q. of certain fish. You must move quickly, panicking the fish only at the last second when the net is scooped from the water. In a stream or river best results are obtained when the net is taken toward the shore. At that last fateful second, the fish are trapped on all sides. The stream bank is along one side, the fishermen and poles are at each end, and the net is along the other side and swooping up underneath. The 6 foot seine is usually best for stream studies because it is easy to handle. It can even be manipulated by one strong person with a pole under each arm, walking with the net held in front.

In a narrow stream some fish can be frightened into a net held stationary. The seine is held in place across the center of the stream. One person stands at each side to operate the poles. A couple of other people must go quietly along the bank to a point 30 to 50 feet upstream and enter the water. Together, they then move downstream quickly, frightening the fish ahead of them by splashing through the water. If the net is held properly, the current should billow it out, shaping it like a bag. By holding the poles at quite an angle, the lower edge of the net can rest on the bottom. With the net in this position, the fish will travel well within the trap. When the people coming downstream are within a yard or so of the net, the seine should be scooped up by those operating the poles. If a second 6 foot seine is available, the people going downstream can prevent any fish from escaping around them. If they walk with the second seine stretched across the stream, nearly all the fish should get trapped in one of the two nets.

For seining in a lake or pond, an area free of plants, such as a beach, is generally necessary. The net, 20–30 feet or more in length, is put in place by boat unless the water is shallow or expert swimmers are available. The seine is brought to shore at moderate speed, forcing the fish into shallow water. To keep the net parallel to the shore, the long ropes used to haul the net in are sometimes marked off in 5 foot lengths. As one group pulls in its side, the marked-off lengths are called out to the other group which tries to stay at the same length. When the ends of the net reach the shore, the fish are trapped as long as the bottom edge of the net rests on the bottom. Although the net is in a large U shape with much still out in the water, it doesn't have to be pulled up on shore. A better idea is to carefully draw the lower edge of the net along the bottom, underneath the trapped fish, and just up on shore. Now all the fish should be writhing on the netting. Two people can work from each end toward the middle, lifting the ends and concentrating the fish in the central section. Representative specimens of various species can then be collected and placed in buckets of water, for identification.

A seine net becomes an "inseine" net when you fail to catch fish even though fish are present in the study area.

6.9 WATER SAMPLING BOTTLE

During studies of lakes, water samples must be collected from several depths. A device called a Kemmerer bottle is effective but expensive. It is recommended that you build a water sampling bottle like the one shown in Figure 6.14.

The water that is collected with this bottle may contain plankton that you should examine. Several bottles of water from the same depth should be poured through a plankton net to get a concentrated sample of the plankton.

You should also perform chemical and physical tests on water samples from various depths. Be sure that the sample is not exposed to air before it is tested, since the dissolved gas content will be changed drastically. For gas content determinations (D.O. and CO_2), a small hose should be fitted to the stop-cock of the Kemmerer bottle. Alternatively, the water can be allowed to flow into the test bottle gently. In addition, for these gas analyses, 3 to 4 times the sample volume should be allowed to overflow so that your sample contains no air bubbles.

Heavy cord (waterproof) marked off in foot or meter intervals

Supporting cord

Rubber stopper

2 liter bottle

Lead sinker attached with epox

Fig. 6.14
A water sampling bott
The bottle is lowered
the required depth w
the stopper in position.
quick jerk removes t
stopper, allowing water
enter the bottle.

6.10 TRANSPORTING TECHNIQUE

Preserved material presents little problem when returning to the school but live material sometimes does. On a long trip, living organisms may suffocate. They can exhaust the oxygen from their holding water unless precautions are taken. A container should *not* be filled to the top; it should have a large surface area where the water sloshing back and forth can get aerated. For transporting fish, special effort is required (Fig. 6.15). Put them in a dark green garbage bag in a bucket with 6 or 8 inches of water in the bag. Inflate the bag with a small oxygen cylinder borrowed from the chemistry department. Oxygen will then easily diffuse into the water en route.

On a long trip, at speeds of 50 m.p.h. or more, an aerator can be operated using a funnel attached to a long piece of rubber tubing (Fig. 6.15). Hold the funnel out the window into the wind and hold the end of the rubber tubing below the surface of the water (¼ of an inch). Air bubbles should be forced into the water continuously.

Fig. 6.15
Two methods of providi
oxygen for living orga
isms being transport
back to the lab.

Green plastic bag

Oxygen

FIELD TRIP TO A POND OR SMALL LAKE

Place

Basically you need a pond or lake which is easily accessible so that a long hike through the brush is avoided. The depth of the water may vary, but it should be at least 5 feet deep in some regions. The size may also vary, but keep in mind that for a larger body of water there will be more factors to consider than for a smaller one. An ideal size is 100 acres or less. Lakes larger than 100 acres should be avoided.

Pre-field Preparation

(a) After choosing the study site, field maps should be prepared. A good map of the area is useful since many observations such as plant and tree species can be recorded directly on it. This gives a permanent record of their locations and abundance.

Maps should be drawn in pencil on good blank paper. Several copies (10 or 15 for a class of 30) should be made of it. Base maps in various scales are usually available from government agencies. The best are of the scale 4 inches = 1 mile or multiples thereof, such as 8 inches = 1 mile or 16 inches = 1 mile. If on the base map the lake is large enough to fill a sheet of paper 8½ x 11 inches, trace it directly. If it is too small, enlargements can be made using a pantograph. Your school geography department probably has one. If you do enlarge a map, however, do not forget to change the scale.

Streams, islands, roadways, railways, dams, and other water control structures should also be shown. In addition, the following should be recorded: the name of the body of water, its geographical position, the elevation, the study date, a "North" sign, and the drawer's name.

(b) Prepare data sheets on which you can systematically record the results of all measurements and tests.

(c) Find out all you can about the past history of the lake or pond. How did it originate? Investigate also the geographical features of the surrounding countryside. Consult a geography or geology teacher for reference materials such as topographical maps and soil maps.

If you are studying a reasonably large lake, you may be able to get helpful information on the lake from the government agency responsible for maintaining water quality. This agency has probably performed for many years the tests that you are going to perform. Why is it important to know all you can about the past history of the lake or pond?

Materials

collecting jars, vials, and buckets

preservative (25% formalin for fish, 70% ethanol for invertebrates)

paper labels and marking pens

notebooks

topographical map

identification guide books

rubber boots and chest waders

rubber gloves

life preservers

boat (2- or 4-man rubber dinghies are safe and easy to work from)

water testing kits (D.O., alkalinity, etc.)

soil testing kit

thermometers

water velocity equipment for inlets and outlets

measuring tape

cord, wire, and weights for depth sounding

water sampling bottle (Kemmerer or homemade)

Secchi disc

Ekman dredge

garden rake

sorting trays and forceps

plankton net

sieves

seine net (6 foot, although for a larger area, nets up to 25 feet can be used)

dip nets

lifters

microscopes: compound and dissecting

microscope slides and eyedroppers for plankton analysis

Objectives

The major objective of this field trip is to determine the relationships that exist between the physical, chemical, and biological factors in the pond or lake.

Procedure

a) Determine the "lay" of the land surrounding the water; is it rolling, hilly, flat? Map prominent features such as inlets, outlets, islands, cliffs, and open beaches. Compare your results with the pre-field map that you prepared.

b) Note the main types of trees and shrubs around the lake or pond.

c) Determine the soil mineral composition, in approximate percentages, of the land at the water's edge. (Soil testing kits contain the necessary instructions.) Do this at several sites around the lake or pond. Of what value is this information? Does the shoreline of the lake or pond consist chiefly of clay, sand, rock, silt, or organically rich soil?

d) Select study sites for the remainder of the work. Each site should be at a spot that appears different from the rest of the body of water. Suggested study sites are:
(1) a site at the main inlet;
(2) a site at the main outlet;
(3) a site where the water is shallow;
(4) a site in a bay (boat may be required);
(5) a site at the deepest spot (boat required);
(6) a mobile site (a boat).

e) Perform the appropriate physical tests at each site: inflow determination at (1); outflow determination at (2); depth profile of entire body of water; transparency determinations by teams in boats; color, turbidity, temperature, and conductivity tests at all sites. The teams in the boats should perform the last 4 tests on water from several depths.

f) Determine the chemical characteristics of the water at each site. As a minimum you should determine the D.O., free carbon dioxide, pH, alkalinity, T.H., T.S.S., and T.D.S. Some of these tests can, of course, be completed back in the laboratory. The teams in boats should perform these tests on water from several depths.

g) Sample the biological organisms at each site. Determine the species present and their relative abundance.
(1) Record the major species of emergent, submergent, and floating plants. Estimate the abundance of each species. Use the following index to record your estimate:

Abundant	A	Occasional	O
Common	C	Rare	R

Note the succession of plant species as you go from the center of the body of water to the shore. What functions do these plants appear to perform in the aquatic ecosystem?

(2) Determine the major species of animals present, using the techniques described in part A of this unit. Be sure to look for animals at each of the locations discussed in Unit 3. Don't overlook the zooplankton.

(3) Determine the main species of phytoplankton at each site.

Notes

a) If the lake or pond is to be revisited, establish a bench mark (B.M.). This is a permanent mark on a cliff face, tree stump, or boulder, from which a measurement is taken to the water surface. In this way water level fluctuations can be detected. Show the position of the B.M. on your map.

b) Do not undertake this study immediately after a heavy rainfall, unless you wish to compare the results with normal conditions.

c) The ideal time for this study is early autumn (September or October). This is still early enough for a thermocline to be present. Also, several species of fish begin their spawning at this time of year. Finally, fewer flies and mosquitoes will bother you. Spring is also a good time to carry out a study. The only drawbacks are that the water is cold and in a constant state of change due, in part, to the spring melt. Also, there is no thermocline. Mid-winter studies should be avoided in colder areas, although many interesting and unusual observations can be made at that time of year.

d) Record the date, time of year, and weather conditions.

e) Look for evidence of man's impact on the pond or lake.

Follow-up

Back in the laboratory you should attempt to analyze the observed differences in physical, chemical, and biological conditions at the chosen locations and depths. Be sure to consider the past history of the lake or pond when you are making your final conclusions. Consider also the impact of man on the lake or pond. In what ways has he modified the terrestrial physical environment? How have these modifications affected the pond or lake? Are there any obvious sources of pollution? How have the pollutants affected the pond or lake?

C. FIELD TRIP TO A STREAM

Place

The basic requirement is an accessible stream where fast water and pools are close together. A number of other considerations should be kept in mind. After heavy rains or during the spring, most waterways become partly flooded, sometimes dangerously so. In the spring, especially, the waters will be cold and possibly murky. An ideal stream width is from 5 to 20 feet, and areas deeper than 3 feet should be avoided. Areas cluttered with fallen limbs will make fishing difficult. A section of a creek downstream from a dam is often rich in animal life and all sorts of plankton. A stream having a few sections with aquatic plants is preferred. A park area will have tables on which to set up equipment such as microscopes.

Pre-field Preparation

Examine a topographical map of the region that includes the stream to be studied. Determine the source or sources of the stream. What kind of countryside does it flow through? How might this affect the water quality? Does it flow through or by a town? How might this affect the water quality?

Materials

collecting jars, vials, and buckets

preservative (25% formalin for fish, 70% ethanol for invertebrates)

paper labels and marking pens

notebooks

topographical map of the area

identification guide books

rubber boots and chest waders

rubber gloves

water testing kits (D.O., alkalinity, etc.)

thermometers

water velocity equipment: Thrupp nails, stopwatch, measuring tape

profile equipment: string, rulers

sieves

plankton net

seine net

dip nets

lifters

sorting trays and forceps

other equipment like a Surber sampler and Ekman dredge, if available

microscopes: compound, dissecting

microscope slides, eyedroppers for plankton analysis

Objectives

The major objective of this field trip is to determine the relationships between the physical, chemical, and biological factors at several different sites along a stream.

Procedure

a) Make a sketch map of the stream. Stand back from the section of the stream under study and try to draw its general shape. Indicate where it widens, narrows, or curves. Show the position of major features like fallen logs, emergent plants, and exposed rocks. Label any other features that can easily be seen from shore, such as rapids, shallow sections, pools, or bottom conditions. This map will be extremely useful when writing a report or preparing a management plan after the field trip.

b) Select the study sites. If possible, a class of 30 students should establish at least 5 sites, spaced at intervals of over 100 feet. If the stream has a mud bottom, the intervals will have to be larger. This is to prevent the activities of one team from disturbing those of a team further downstream. Try to pick study sites with differing environmental conditions. Suggested study sites are:
(1) a site in a riffle area (with a rock or gravel bottom);
(2) a site with a sandy bottom;
(3) a site with a mud or silt bottom;
(4) a site with a bedrock bottom;
(5) a site near the shore (shallow water);
(6) a site near the center (deep water);
(7) a site with dense vegetation;
(8) a site with little or no vegetation.

Mark the chosen sites on your map.

c) At each site carry out a study of the physical characteristics of the stream: cross-sectional profile, velocity of flow, volume of flow, temperature, color, turbidity, and conductivity.

d) At each site do a chemical study of the water. As a minimum you should determine the D.O., free carbon dioxide, pH, alkalinity, T.H., T.S.S., and T.D.S. Some of these tests can, of course, be completed back in the laboratory.

e) At each site, sample the biological community to determine the species present and their relative abundance. Record the abundance as follows:

| Abundant | A | Occasional | O |
| Common | C | Rare | R |

Check Unit 4 again to see where you should look for the various organisms. Use the sampling techniques described in part A of this unit.

Notes

a) Keep in mind that some activities in the stream will disrupt other planned activities. The chemical tests should, in general, be done before the water and bottom silts are disturbed. Plankton samples should be taken early. The invertebrate collections should be completed and properly preserved, stored, and labeled before seining and other fish collecting is even attempted. Physical factors other than color and turbidity could be analyzed last.

b) Make a note of the date, time of day, and weather conditions.

c) Microhabitats within each study site should be examined. For example, compare the organisms on the upstream face of a rock with those on the downstream face of the same rock. Or, compare the organisms in a "riffle" area with those in a nearby "pool" area.

d) If possible, the same sites should be studied at another time of year. Physical factors like temperature and water velocity normally vary with the seasons. As a result, chemical factors like D.O. also vary. Changes in the physical and chemical factors produce changes in the distribution, abundance, and, perhaps, in the types of organisms present.

Follow-up

Back in the laboratory you should attempt to analyze the observed differences in physical, chemical, and biological conditions at the various study sites. When you are making your conclusions, be sure to consider the source of the stream, the type of countryside through which it flows, and man's effects on the stream (the presence upstream of towns or industries).

If conditions of the stream need improvement, either above or below the town, a management plan in the form of a map could be drawn to show the position and the type of improvement that you think would be helpful. You may wish to submit this plan and a report of your findings to the town council and to any government agency that is responsible for the maintenance of water quality in that particular waterway.

Case Studies

7

The following case studies consist of actual information collected by scientific means. They are included to give you a chance to find out if you can apply the knowledge that you have gained from this book. By now you should be quite aware of the interdependence of biotic (living) and abiotic (physical and chemical) factors in an ecosystem. If any one of these factors is altered, most of the others also change. If, for example, the temperature of a body of water is changed, the D.O. changes and living organisms are affected. If first-order carnivores are harmed in some way, herbivores may increase and, consequently, producers may decrease. This, in turn, affects the D.O. The chain of relationships is unending, complex but certain. You have undoubtedly discovered this in your field and laboratory work.

Try these case studies to see how well you understand the "ecosystem concept."

Case Study 1 STREAM IMPROVEMENT

Trout are highly prized by fishermen. Unfortunately, they are extremely intolerant fish and quickly succumb to mismanagement and disruption of our lakes and streams. Wildlife managers have developed a number of practices to counteract harmful conditions in streams. These practices make the environmental conditions more suitable to trout. In some cases they can produce trout water where none existed before.

The ecological requirements of trout are basically:

a)   cool water;

b)   well-oxygenated water;

c)   sections of gravel bottom (for spawning);

d)   clear water (for catching food);

e)   occasional pools (where the fish feed);

f)   adequate food (aquatic and terrestrial life, the latter usually falling from surrounding vegetation).

Figure 7.1 shows some management techniques used to improve the conditions in a hypothetical stream. Every one of the requirements of trout has been improved in one way or another. List each of the techniques and describe fully how each will benefit the trout.

. 7.1
per management can
rove conditions in a
am.

Stone deflector

Single wing deflector
(log buried in bank)

Pool

Current

Gabian baskets
(baskets made of wire
containing rocks)

Previous
erosion

Willow
and
white cedar
plantings

Shade
plantings

Rip-rap (large
boulders along
shore)

Drop dam
(water pours
over top)

Pool

Gravel

Pool

Digger log
(forces water
underneath)

Planting

Double wing
deflectors

Boulder retards

155

TABLE 3

Station	Latitude	Longitude	Days	Total Depth (meters)	Depths (meters)	Temp. (°C)	Turbidity (units)	Oxygen Winkler (mg O_2/l)	Coliforms per 100 ml
A	43°39'N	78°57'W	10/2/67	117	1	12.08	0.3	—	0
	Lake Ontario				10	10.06	0.4	—	0
					20	8.93	0.3	—	—
					30	5.69	0.2	—	0
					50	4.02	0.2	—	—
					75	3.88	0.2	—	—
					100	3.79	0.2	—	—
					115	3.78	0.3	—	0
B	43°22'N	77°30'W	10/3/67	121	1	15.20	0.2	—	0
	Lake Ontario				10	15.04	0.2	—	0
					20	14.55	0.1	—	—
					30	14.20	0.2	—	0
					50	4.10	0.2	—	—
					75	3.93	0.1	—	—
					100	3.90	0.4	—	—
					119	3.84	0.1	—	0
C	43°47'N	76°37'W	10/4/67	63	1	16.06	0.6	—	—
	Lake Ontario				10	16.08	0.6	—	—
					20	15.83	0.6	—	—
					30	12.47	0.5	—	—
					50	4.50	1.1	—	—
					61	4.46	1.2	—	—
D_1	42°31'N	79°53'W	10/4/67	64	1	15.51	1.1	9.83	0
	Lake Erie				10	15.40	1.2	9.61	—
					19	15.33	1.2	9.44	—
					31	13.54	1.2	8.95	—
					49	5.26	1.4	8.42	—
					61	5.17	2.1	8.21	0

Station	Latitude	Longitude	Date		Depth				
D₂	42°31'N	79°53'W	6/2/67	64	1	13.41	1.0	13.54	0
	Lake Erie				4	9.01	2.2	15.05	0
					7	8.17	2.6	13.25	—
					10	7.38	2.2	12.58	—
					19	6.22	1.0	12.73	—
					22	4.97	1.6	12.63	—
					49	4.54	1.5	12.40	0
					61	4.32	2.8	12.13	
D₃	42°31'N	79°53'W	7/12/67	63.7	1	19.79	—	—	—
	Lake Erie				7	19.73	—	—	—
					10	18.05	—	—	—
					13	15.02	—	—	—
					16	8.55	—	—	—
					31	5.16	—	—	—
					61	4.40	—	—	—
E	41°55'N	83°04'W	7/17/67	9.4	1	20.54	3.1	8.50	14
					4	19.46	3.1	8.39	—
					7	19.25	2.5	7.79	67
F	41°42'N	83°27'W	7/18/67	9.8	1	—	12.5	6.02	800
					4	—	12.0	6.13	—
					7	—	11.0	6.00	200
G	41°43'N	82°45'W	7/18/67	10.7	1	21.58	3.2	7.38	3
					4	21.67	3.1	7.64	—
					7	21.67	3.2	7.54	1

MONITORING LAKE ONTARIO AND LAKE ERIE

Various government and private institutions are currently involved in monitoring many lakes throughout the world with the hope of understanding what is happening to our large and small bodies of water. Some of the recent concern over man's degradation of his environment has prompted this interest. For a number of years, Lake Ontario and Lake Erie have been intensively studied by the Canada Centre for Inland Waters. A great deal of data has been collected on many chemical, physical and biological features. A small sample of the water quality data is shown in Table 3 on pp. 156–157. These records are for water sampling stations that you can locate on a map using their directional coordinates. The day, month, and year that the data were obtained are also listed.

It is possible to graphically represent some of these data to make them more meaningful. On one graph, plot the three sets of temperature data for Lake Ontario, all gathered within a three day period. On the graph, make the horizontal "x" scale the temperature variable, and the vertical "y" scale the depth variable. Was a thermocline evident at all three stations? The prevailing wind that sweeps over Lake Ontario comes from the west. Could this have any relevance to the depth of the warm surface waters at various points? Could this fact explain the differences observed? If you have ever gone swimming in a large lake when there was a strong off-shore breeze you couldn't mistake this phenomenon.

In a similar fashion, plot the temperature values for the first three sets of data for Lake Erie on a separate graph. Note that these data pertain to a given sampling station at three different times in the year, early June, mid-July, and mid-October. Does the thermocline shift in depth or configuration? Although no extreme oxygen deficit occurs at any depth, does there appear to be a pattern in the data? If so why? What might be in the water in early June (station D_2) that gives the generally larger turbidity readings at this time?

Stations E, F, and G are found in western Lake Erie, a rather shallow area. What might account for the turbidity and oxygen values obtained here? Do you think it significant that station F, located in Maumee Bay, near Toledo, Ohio, has coliform counts that vary markedly from the other stations in Lake Erie and Lake Ontario?

Select the two most comparable sets of data for Lake Erie and Lake Ontario and discuss the possible significance of the difference in turbidity readings.

mmon me	12 hr. TLm (°C)
mmon shiner	32.0
ook trout	25.5
arl dace	31.1
rp	41.0
rgemouth bass	36.4
mpkinseed	34.5
uegill sunfish	34.3
acknose dace	29.4
ellow perch	30.8
own bullhead	34.8
olden shiner	34.7
athead minnow	33.7
eek chub	31.5
orthern redbelly ace	33.1

Case Study 3 LETHAL TEMPERATURE EXPERIMENTS

Through laboratory experiments, ecologists have gained much of their knowledge concerning the tolerance of animals to variation in their environment. One measurement that shows tolerance is called the *24 hr. TLm (tolerance limit median)*. What does this mean? Let us look at an example. A researcher obtains a collection of similar sized largemouth bass and wants to find their tolerance to various temperatures. His studies eventually lead to the conclusion that one half or 50% of the fish placed in water at 29°C die during the first 24 hours. The 24 hr. TLm is 29°C.

At a higher or more lethal temperature than 29°C, more than 50% of the fish would die after 24 hours. In fact, at some higher temperature, half of the fish might already have perished in 12 hours. You would call this higher temperature a *12 hr. TLm*. The 12 and 24 hour experiments are the main ones carried out on fish. If you understand TLm you should be able to express in words what a 36.5°C, 12 hr. TLm signifies. Can you?

The accompanying table lists some fish species often found in streams and rivers, with their 12 hr. TLm values. (These values are hypothetical but are believed by the authors to closely approximate the true values.) It is possible, although improbable, that these species could live together in one river.

Suppose an industry along the river poured very hot water into the stream for a day or so, drastically raising the temperature of the water. Which would be the first five species to die from thermal pollution? Which five species would be most tolerant?

If the water contained toxic cyanides that gave the river water a concentration of 0.03 mg per liter, do you think the same five species would be the first to die? If you wished to know more about the tolerance of the various species to cyanides, how would you set up an experiment to reveal this?

Case Study 4 LAKE TELETSKOE, U.S.S.R.

Figure 7.2 shows the composition of some typical communities of bottom animals dredged from a variety of depths. Although these data pertain to Lake Teletskoe in the U.S.S.R., the ecology of lakes is similar throughout the world. To make these data more meaningful, estimate the percentage of each type of organism at each depth. Plot the percentages graphically against depth. Or, in your own words, describe any trends that you can observe in the types of animals inhabiting the bottom as the water gets deeper. Why do these trends exist?

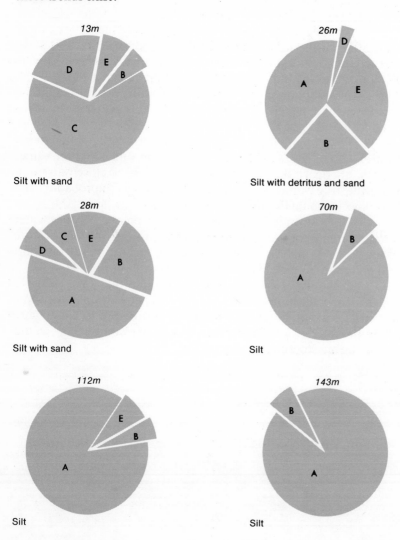

Fig. 7.2
Proportions of bottom dwelling organisms various depths in La Teletskoe. (A) oligo chaetes, (B) midge larva (C) bryozoans, (D) clan (E) crustaceans.

Case Study 5   FOLSOM DAM

Streams and rivers vary throughout the year in the amount of water which they carry. Part of this variation is predictable and depends on the climatic characteristics of the specific region in which a river is located. In central North America (Nebraska, the Dakotas, and south Saskatchewan), precipitation comes primarily through rain in the summer. In contrast, California invariably suffers from a summer scarcity of precipitation. Its daily changes in weather phenomena are somewhat less predictable. In a few hours a single storm can leave behind an inch or two of precipitation. The result is a widely varying flow of water in practically every natural stream or river.

Dams have been built for many purposes, including hydro-electric power, irrigation, and flood control. This latter use is significant in many of the northern states and Canada, where a large portion of the run-off produced as spring melt-water is trapped and stored in the reservoirs. This water is released over a long period of time, especially during drier months when the river flow might otherwise dwindle to nothing. What is often overlooked in such projects is the tremendous change in conditions and the ecology of the river downstream from a dam. This is something that you should now be able to predict.

Records were kept for many years on the discharge of water along the American River in California. The data were obtained before and after the Folsom Dam was constructed in 1955 at Fair Oaks. The discharge was measured in terms of cubic feet per second (cfs). The records kept tab on the number of days the discharge was within given rate limits. In Table 5 on p. 162 you should recognize quite a difference in the pattern of discharge before and after the dam was established. Calculate the percentage of the days when the discharge was:

a)	less than 500 cfs;
b)	more than 500 but less than 10,000 cfs;
c)	more than 10,000 cfs;

both before and after the dam was built. You should be able to discuss the significance of these percentages for the ecology of the river.

How might the physical nature of the stream bed change with the absence of periodic torrential discharges? How could well-oxygenated water be virtually assured after construction of the dam? How would the communities downstream be affected?

TABLE 5

Rate of Discharge (cfs)	Number of days discharge was within rate limits	
	Before Dam	After Dam
0 – 5	16	0
5 – 10	15	0
10 – 20	24	0
20 – 30	17	0
30 – 50	36	0
50 – 100	297	0
100 – 200	987	0
200 – 300	1,363	0
300 – 500	2,093	2
500 – 1,000	2,673	232
1,000 – 2,000	2,062	484
2,000 – 5,000	2,902	776
5,000 – 10,000	2,752	223
10,000 – 20,000	1,319	80
20,000 – 30,000	151	18
30,000 – 50,000	56	8
50,000 – 70,000	10	4
70,000 – 100,000	11	0
100,000 – 150,000	3	0
Total Days	16,787	1,827

When a plant responds to a stimulus, the response is called a *trop-ism*. In particular, a response to light is called phototropism. The response of an animal to a stimulus is called a *taxis*. An animal that moves towards a light source is called positively phototactic. An animal that moves away from a light source is called negatively phototactic. There are other forms of taxis, for instance, geotaxis is a reaction to gravity, and thigmotaxis is a response to contact.

Final positioning of an organism in its environment depends to a certain extent on its various tactic responses. Some organisms react to the presence or absence of members of their own species, that is, they may or may not be sociable (gregarious). Crayfish are positively thigmotactic, but non-sociable animals. They tend to space themselves out away from one another even where few hiding places are available. When they are placed in a container, they often end up in the corners or around the edge. The extra contact gained in a 90° corner is at least better than standing in the center, with contact only from below. Mayflies taken from a stream and placed in a glass container filled with water but no substrate, invariably cluster together. They seem to desire something to cling to even if it has to be the body of another organism. In contrast to crayfish, they are not antagonistic.

g. 7.3A
nimal behavior. Distribu-
on of 2 species of organ-
ms in 2 containers of
milar shape and size.

6	3	0	0	2	0	0	0	5	9
6	0	0	0	0	0	0	0	0	3
2	0	0	0	0	0	0	0	0	2
1	0	0	0	0	0	0	0	0	0
0	0	0	0	0	0	0	0	0	0
0	0	0	0	0	0	0	0	0	0
0	0	0	0	0	0	0	0	0	4
0	0	0	0	0	0	0	0	0	0
5	0	0	0	0	0	0	0	0	4
8	3	0	4	0	0	0	0	1	7

4	1	1	2	1	1	2	1	1	3
2	1	0	0	0	1	0	0	1	1
1	0	0	1	0	0	1	0	0	1
2	0	0	0	0	0	0	0	0	3
1	0	0	0	0	0	0	0	1	1
1	1	0	0	0	0	0	0	1	1
3	0	1	0	0	0	1	0	0	2
1	2	1	0	0	0	0	0	1	1
1	1	0	1	0	0	1	1	2	1
4	1	1	1	1	2	1	1	1	2

Figure 7.3 A shows the number of organisms of two different species collected in samples from corresponding positions in two similar sized and shaped containers. Both of the species showed positive thigmotactic responses. In one case the animals were antagonistic to one another, whereas in the other case they were gregarious. First look at the figure to see how the animals were distributed in the containers. For each species, count the

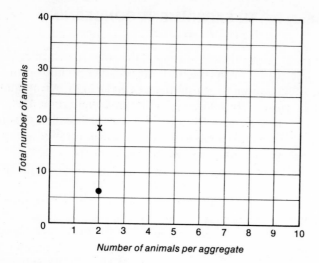

Fig. 7.3B
A sample graph.

number of samples in which no organisms were found. Each container had 75 organisms in it. By counting the number of empty samples, can you already predict which was the gregarious species?

Similarly, for each species, count the total number of animals found in samples with only one animal present, then 2, 3, 4, 5, etc. You might try plotting these data on a graph as shown in Figure 7.3 B. Two of the tabulations have already been plotted. They show that for one species there were 6 animals that aggregated in 2's while for the other species 18 animals aggregated in 2's. Ignore the number of samples with no animals on the graph. Does the graph show the gregarious nature of one of the species?

How do you know the antagonistic species was thigmotactic in the first place? Do other kinds of responses (to particular stimuli) sometimes inhibit thigmotaxis in the crayfish?

Case Study 7 TEMPERATURE PROFILES

The sampling stations on Lake Ontario, used by the Canada Centre for Inland Studies, are not haphazardly selected. For example, eight of the stations used in 1967 were placed in a direct line across the lake. The stations were arranged in this manner to obtain a profile or cross-section of conditions in the lake.

If you had been aboard the vessel Theron in 1967, you would have helped to collect the temperature data shown in the accompanying table. The ship went on two cruises and, as you can

TABLE 6

Sampling Station	1	2	3	4	5	6	7	8
Location Lat. Long.	43°-54' 77°-30'	43°-48' 77°-30'	43°-43' 77°-30'	43°-38' 77°-30'	43°-33' 77°-30'	43°-27' 77°-30'	43°-22' 77°-30'	43°-17' 77°-30'
Sounding (meters)	30.5	53.5	78.0	114.0	167.5	163.5	114.0	30.5

Temperatures on July 27, 1967 (°C)

Depth	1	2	3	4	5	6	7	8
1	16.0	19.5	19.5	20.0	20.0	21.5	20.5	21.5
10	9.0	7.5	19.5	20.0	8.5	21.5	20.0	21.5
20	8.5	6.0	5.0	7.5	5.0	10.0	16.5	21.5
30	(28)8.0	5.0	4.5	4.5	4.5	5.5	8.5	21.0
50		4.5	4.5	4.0	4.0	4.0		4.5
75			4.0	4.0	4.0	4.0		7.5
100				4.0	4.0	4.0		4.0
150				(110)4.0	4.0	4.0		(110)4.0
					(165)4.0			(160)4.0

(continued)

Table 6 (continued)

Sampling Station	1	2	3	4	5	6	7	8
	Temperatures in early November, 1967 (°C)							

Depth (meters)	1	2	3	4	5	6	7	8
1	9.5	10.0	8.0	7.5	8.0	9.5	9.5	1(
10	9.5	10.0	8.0	7.5	8.0	9.5	9.5	10
20	9.5	10.0	8.0	7.0	8.0	9.5	9.5	10
30	(28)9.5	10.0	8.0	6.5	8.0	9.5	9.5	10
50		7.5	8.0	8.5	8.0	4.5		9
75			5.0	5.0	7.0	4.0		7
100				4.0	4.5	4.0		4
150				(110)4.0	4.0			(110)4
					(165)4.0			(160)4

see, it returned to many of the same sampling points to repeat measurements at different times. As a member of the crew, you might have been called upon to make a temperature profile of the lake for the two different cruises. Do this, and then draw isotherms (lines of similar temperature) for 4.5°C, 6.0°C, 8.5°C, and 19.0°C.

Try to account for the differences in the two profiles. Why does there appear to be a greater build-up of warm water on the southern side of the lake in July than on the northern side?

Case Study 8   AN EXAMPLE OF pH VARIABILITY

Four streams in South Wales were once examined for their chemical properties. The techniques used were comparable to those in your field studies. Some interesting results concerning hydrogen ion concentration were obtained which shed light on the variability of pH in running water.

TABLE 7

| River | Flood level | pH | Hardness (mg/liter) | |
			Ca	Mg
Amman	Normal	5.2	2.8	1.2
	Full flood	4.4	2.2	0.5
Garw	Normal	6.4	1.8	2.0
	Full flood	6.0	–	1.0
Pedol	Normal	5.8	3.2	2.0
	Full flood	4.4	2.4	1.5
Clydach	Normal	7.8	25.8	5.0
	Full flood	6.8	6.5	–

Three of the four streams, Amman, Pedol, and Garw, were found to contain soft water. Clydach contained hard water. The chemical data for the four streams are shown in Table 7. (A dash means no determination was made.) As you can see, data were obtained under two different conditions, normal flow and full flood. In every case, flood conditions gave lower, more acidic, pH conditions. After lengthy investigation, a definite pattern was observed between the amount of flow and the pH. The data from more than 60 determinations taken over half a year are summarized in Table 8. Transfer these data to a graph. On the "y" axis, place the pH values varying from 4.0 to 8.0 from top to bottom. On the "x" axis, space out from left to right the arbitrary units for water flow: "low," "normal," "slight flood," "medium flood," and "full flood."

Why do the pH values vary from stream to stream and at different degrees of flood? The answer lies in the source of the water. A few hints should lead you to the answers.

a)   The Clydach is fed mainly by spring water when at normal flow.

TABLE 8

Stream level	pH Values			
	Amman	Garw	Pedol	Clydach
Low	5.7	6.8	6.0	8.0
Normal	5.1	6.4	5.8	7.8
Slight flood	4.8	6.4	5.6	7.6
Medium flood	4.5	6.2	5.1	7.2
Full flood	4.3	6.0	4.4	6.8

b) Acids added to hard water containing calcium bicarbonate and magnesium bicarbonate are neutralized as in the reaction

$$Ca(HCO_3)_2 + 2HCl \rightarrow CaCl_2 + 2H_2O + 2CO_2$$

c) Much of the Welsh landscape at high altitudes, such as in the area of these streams, has very peaty soils. (Peaty soil contains a great deal of organic matter.)

Case Study 9   A DEADLY MATHEMATICAL QUESTION

Lake Erie, the smallest by volume of the Great Lakes, has felt the effects of man's polluting ways the heaviest of all. Even if pollution ceased overnight and the lake had a chance to replace the present water, this cleansing process would take a number of years. By calculating the volume (V) to outflow (O) ratio, you can get some idea of the time that would be involved. You must realize, however, that through the various mixing processes within the lake, some soluble pollutants may take a devious route (much slower than the V/O value would suggest) before reaching the Niagara River, the outflow point. The volume of Lake Erie is 458 km³ and the rate of outflow is 175 km³/year. The resident time (RT) of water soluble substances is V/O or slightly less than 3 years. Similar calculations can be made for the other lakes by consulting the following table.

TABLE 9

Lake	Volume (km³) (V)	Outflow (km³ per year) (O)	Resident Time (RT)
Erie	458	175	2.6 Years
Superior	12,221	65	
Michigan	4,871	49	
Huron	3,535	159	
Ontario	1,636	208	

Suppose that 1,000 gm of soluble radioactive waste were spilled into Lake Superior at Thunder Bay. Eventually this material would flow through Lakes Superior, Huron, Erie, and Ontario. It would then be carried by the St. Lawrence River to the Atlantic Ocean. The time spent by the material in each lake is the RT value for each lake. Thus the total time required for the material to pass through the Great Lakes system can be found by totalling the RT values. Since radioactive material loses its radioactivity as it decays, not all 1,000 gm will be radioactive when the waste reaches the outflow point of the Great Lakes. If the half-life of the radioactive material is 110 years, how many grams of the material will still be radioactive by the time it reaches the outflow point? (The half-life of a radioactive substance is the time required for the substance to lose half of its radioactivity.) Locate Montreal, Quebec, on a map. Where do you think this large city gets its drinking water? How much radioactive material do you think will eventually reach Montreal? Can you think of some of the reasons why at least a portion of the soluble waste in Lake Superior would remain for thousands of years?

Case Study 10   MANAGING A RESOURCE

Small lakes and ponds can be put to numerous uses by man. These uses include sport fishing, swimming, boating, waterfowl hunting in the fall, bird watching, sanctuaries for waterfowl production, and minnow (bait fish) production. Fish Lake is a shallow body of water less than a mile in length, found in Prince Edward County, the island-like land mass located along the northern edge of Lake Ontario. The small lake was once surveyed to determine suitable uses to which it could be put. The recommendations arising from the survey were:

a)   that the lake be set aside as a waterfowl sanctuary;

b)   and that the lake be established as a source for bait minnows, the rights to fishing being leased to bait sellers that supply to the summer tourist trade.

The establishment of the bait industry was based on the fact that only six species of fish were present in the lake. The golden shiner and brown bullhead were the most abundant. The former fish is known to be an excellent bait minnow, since it grows rapidly and is quite tolerant to handling. It was recommended that, first, all fish should be removed from the lake using Rotenone (a chemical toxic to fish). Then, after the Rotenone concentration had dropped below the toxic level, golden shiners should be reintroduced as the only fish species.

Extensive die-offs of fish had occurred in the past in this lake, during the winter. Why might this have happened? Could problems develop as a result of having only one fish species in the lake? Would the proposed bait industry be more or less susceptible to winter die-offs or is this predictable?

The accompanying map (Fig. 7.4) was made of the vegetation of the lake. The wild rice, *Zizania*; narrow-leafed pond weeds, *Potamogeton pectinatus*; and wild celery, *Vallisneria*, are good duck foods for the establishment of the waterfowl sanctuary. The dominant plant for most of the area was *Chara*, a stonewort, a very unsuitable wildlife plant.

What proposals could be made for alteration of the aquatic vegetation to produce a better habitat for waterfowl? How would this be accomplished? Can a lake be cultivated to produce certain foods for wildlife? What other considerations must be taken concerning the "needs" of ducks and other waterfowl? Since ducks and especially ducklings feed a great deal on aquatic invertebrates, do you think the two proposed programs, minnow and waterfowl production, might interfere ecologically with one another? Why?

LEGEND

CHIEF VEGETATION

Decodon
Lysimachia
Myriophyllum
Nuphar
Nymphaea
Potamogeton pectinatus
Potamogeton natans
Pontederia
Sagittaria
Scirpus
Typha
Vallisneria
Zizania

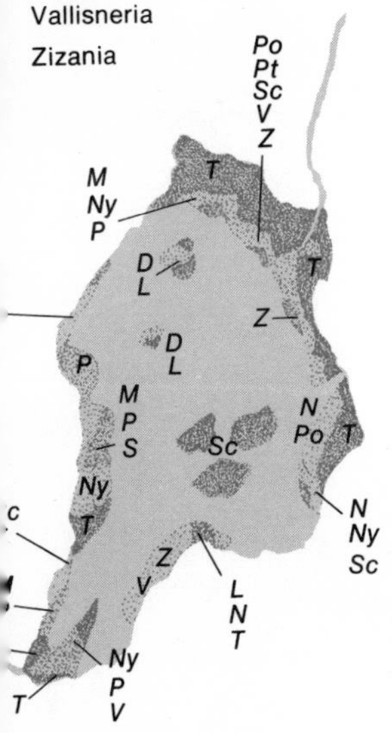

Fig. 7.4
Chief vegetation of Fish Lake. A mat of *Chara* covers most of the remaining bottom of the lake.

Case Study 11   LAMPREYS IN THE GREAT LAKES

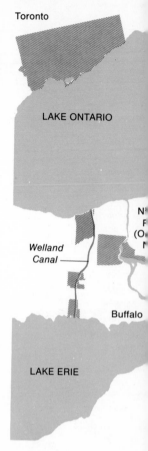

The sea lamprey feeds by attaching itself to the body of a fish with its funnel-like mouth opening. Then it pierces the skin with its tooth-lined mouth and draws out blood and other body fluids. This leaves a raw sore on the fish, which can become infected, eventually causing death.

In 1921, the sea lamprey entered the upper Great Lakes through the newly opened Welland Canal (Fig. 7.5). The declines in the catches of the lake trout fisheries were as follows:

Lake Huron: 5,000,000 lb in 1938 to 400,000 lb in 1947
Lake Michigan: 6,500,000 lb in 1944 to 400 lb in 1953
Lake Superior: 4,400,000 lb in 1947 to 2,200,000 lb in 1956

How do you account for the decrease in lake trout production as you go from near the Welland Canal to points farther away? How do you account for the fantastic drop that occurred in Lake Michigan lake trout production, as compared to the drops in the other two lakes?

Lake trout mature at about 7 to 8 years. The hybrid splake (or wendigo) matures at 3 years. It is felt that the splake could replace the lake trout in the above lakes. What characteristics must the splake have in order to occupy the niche left by the lake trout, as well as contending with the lamprey, assuming that lampreys are still present?

How do you think a tragedy such as this could have been prevented if scientists had known that the lamprey existed in Lake Ontario prior to 1921?

Fig. 7.5
Location of the Welland canal.

Case Study 12   A LAKE SURVEY

The data in Table 10 are from a lake that was surveyed in 1968.

TABLE 10

Depth (feet)	D.O. (ppm)	Alkalinity (ppm)	Temperature (°F)
0	8	34.2	67
5	–	–	67
10	8	34.2	66.5
15	–	–	66.5
20	9	34.2	66
25	–	–	66
30	7	34.2	65
35	–	–	62
40	9	34.2	56
45	–	–	53
50	8	34.2	51
55	–	–	50
60	8	34.2	49
70	8	–	49
80	8	34.2	49
90	6	–	48
100	6	34.2	47
110	6	–	47
120	3	34.2	46

T.H. at 10 feet — 34.2 ppm
T.D.S. at 10 feet — 30 ppm
Turbidity at 10 feet — 6 J.T.U.
Conductivity at 10 feet — 45 μmho
Secchi disc reading — 13.5 feet

Fish found: lake herring, lake whitefish, yellow perch, white sucker, lake trout, cisco, catfish, burbot.

Plants found: pondweed, Canada water weed, water milfoil, yellow water lilies, white water lilies, bulrushes.

1)	Plot a depth versus D.O. and temperature graph. In what season was this study done?
2)	At what depths would you expect to find the greatest number of trout-like fish?
3)	How do you account for the large amount of oxygen dissolved in water between 40 and 100 feet?
4)	Where is the thermocline?
5)	Based on the alkalinity and T.H. figures, what type of bedrock would you expect to find around this lake?
6)	If the yellow perch population were quite large, what would you expect the populations of each of the other fish species to be? Why?
7)	Discuss the nature of this lake with respect to species of plants and animals present and the chemical and physical environment in which they live.

Case Study 13   FOOD CHAINS AND FOOD WEBS

In a very productive pool in Bennet's Stream in Manitoba, you are likely to find the following organisms:

planaria
diatoms
brown algae
copepods
cladocerans
mayfly nymphs
stonefly nymphs
dragonfly nymphs
midge larvae
crayfish
trout
shiners
dace
nematodes
water pennies (riffle beetles)
predaceous diving beetles
water striders
caddisfly larvae

Surrounding the pool are many large vertebrates. These include muskrats, raccoons, snakes, frogs, rodents, and several species of hawk. In the pool fringes, some higher plants such as bur-reeds and arrowheads are abundant.

1)   List ten food chains involving any of the above organisms.

2)   Construct a food web for the pool. Compare your results with those of other members of your class.

WOOD CHIPS IN AN AQUATIC ENVIRONMENT

A river in British Columbia was examined at two stations, A and B. At station A, the following organisms were found in abundance:

> mayfly nymphs
> stonefly nymphs
> caddisworms
> water pennies (riffle beetles)
> diatoms

At station B, the following were abundant:

> *Tubifex*
> bloodworm (burrowing midge larvae)
> blue-green algae
> dragonfly nymphs
> nematodes

Between stations A and B, a lumber mill dumped a large quantity of wood chips into the surrounding water.

---

1) What important chemical change from station A above the mill and station B below the mill, is indicated by the change of organisms?

---

2) What could have caused this change to occur?

---

3) Name 3 types of fish which you would expect to find at station A and at station B.

---

4) Assume that the lumber mill ceases to dump wood chips into the water. What steps could be taken to hasten the return of the environment at station B to the conditions prevailing at station A?

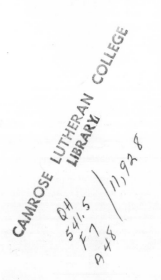